Dynamics of Neural Networks

Michel J. A. M. van Putten

Dynamics of Neural Networks

A Mathematical and Clinical Approach

Springer

Michel J. A. M. van Putten
Clinical Neurophysiology Group
University of Twente
Enschede, The Netherlands

Neurocenter, Dept of Neurophysiology
Medisch Spectrum Twente
Enschede, The Netherlands

ISBN 978-3-662-61182-1 ISBN 978-3-662-61184-5 (eBook)
https://doi.org/10.1007/978-3-662-61184-5

This Springer imprint is published by the registered company Springer-Verlag GmbH, DE part of Springer Nature.
The registered company address is: Heidelberger Platz 3, 14197 Berlin, Germany

Preface

This book evolved from the course "Dynamics of Neural Networks in Health and Disease." It treats essentials from neurophysiology (Hodgkin-Huxley equations, synaptic transmission, prototype networks of neurons) and related mathematical concepts (dimensionality reductions, equilibria, bifurcations, limit cycles, and phase plane analysis). This is subsequently applied in a clinical context, focusing on EEG generation, ischaemia, epilepsy, and neurostimulation.

The book is based on a graduate course taught by clinicians and mathematicians at the Institute of Technical Medicine at the University of Twente. Throughout the text, we present examples of neurological disorders in relation to applied mathematics to assist in disclosing various fundamental properties of the clinical reality at hand. Exercises are provided at the end of each chapter; answers are included.

Basic knowledge of calculus, linear algebra, differential equations, and familiarity with Matlab or Python is assumed. Also, students should have basic knowledge about essentials of (clinical) neurophysiology, although most concepts are shortly summarized in the first chapters. The audience includes advanced undergraduate or graduate students in Biomedical Engineering, Technical Medicine, and Biology. Applied mathematicians may find pleasure in learning about the neurophysiology and clinical applications. In addition, clinicians with an interest in dynamics of neural networks may find this book useful.

The chapter that treats the meanfield approach to the human EEG, Chap. 7, was written by Dr. Rikkert Hindriks. The Chaps. 3 and 4, discussing essentials of dynamics, were in part based on lecture notes by Prof. Stephan van Gils, Dr. Hil Meijer, and Dr. Monica Frega made various useful suggestions to previous versions. Further, Annemijn Jonkman, Bas-Jan Zandt, Sid Visser, Koen Dijkstra, Manu Kalia, and Jelbrich Sieswerda are acknowledged for their critical reading and commenting on earlier versions. Finally, I would like to thank our students and teaching assistants who provided relevant feedback during the course.

Enschede, The Netherlands | Michel J. A. M. van Putten

The original version of the book was revised. The correction to this book is available at https://doi.org/10.1007/978-3-662-61184-5_11

Prologue

> How can a three-pound mass of jelly that you can hold in your palm imagine angels, contemplate the meaning of infinity, and even question its own place in the cosmos? Especially awe inspiring is the fact that any single brain, including yours, is made up of atoms that were forged in the hearts of countless, far-flung stars billions of years ago. These particles drifted for eons and light-years until gravity and change brought them together here, now. These atoms now form a conglomerate—your brain—that can not only ponder the very stars that gave its birth but can also think about its own ability to think and wonder about its own ability to wonder. With the arrival of humans, it has been said, the universe has suddenly become conscious of itself. This, truly, is the greatest mystery of all.
> —V. S. Ramachandran, The Tell-Tale Brain: A Neuroscientist's Quest for What Makes Us Human

A 64-year old, previously healthy, patient was seen at the emergency department. He woke up this morning with loss of muscle strength in his left arm and leg. On neurological examination, he has a left-sided paralysis. A CT-scan of his brain showed a hypodensity in the right middle cerebral artery area, with a minimal shift of brain structures to the left, characteristic for a cerebral infarct (Fig. 1). His wife tells you that he complained about some loss of muscle strength already the evening before. He is admitted to the stroke unit. Two days later, he is comatose, with a one-sided dilated pupil. A second CT-scan shows massive cerebral edema of the right hemisphere with compression of the left hemisphere and beginning herniation. The day after, he dies. What happened? Why did his brain swell? Which processes are involved here? Could this scenario have been predicted and perhaps even prevented?

A 34-year old woman is seen by a neurologist because of recurring episodes of inability to "find the right words." These episodes of dysphasia recur with a variable duration and frequency, sometimes even several times per day. The duration is up to several minutes, and recovery takes up to half an hour. She suffered from a traumatic brain injury half a year earlier and on her MRI scan a minor residual lesion was shown near her left temporal lobe. Despite treatment with various anti-epileptic drugs, she does not become seizure free. Early warning signs are essentially absent and she finds it difficult to continue her job as a high school teacher. Why did her

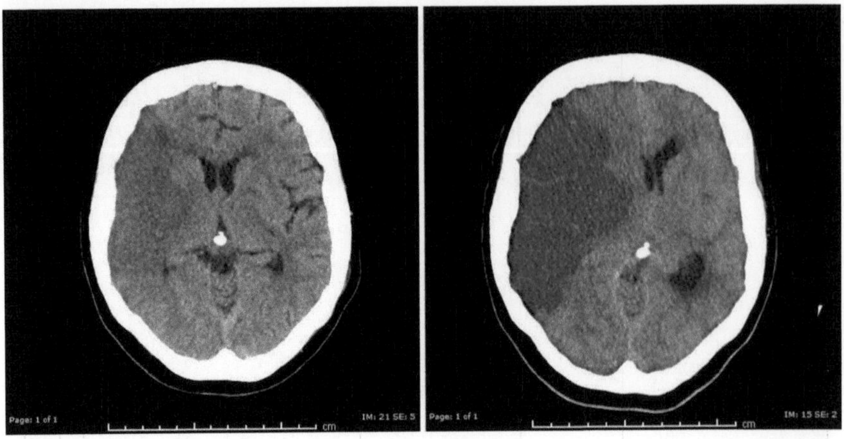

Fig. 1 Left: Noncontrast head CT of a patient with an acute right middle cerebral artery infarction, showing hypodense gray and white matter on the right side of the brain. Note that this is left in the image, as we "look from the feet upwards to the head" of the patient. Right: head CT two days later shows an increase in the hypodensity and marked swelling of the infarcted tissue on the right, with significant cerebral edema and brain herniations. Courtesy: M. Hazewinkel, radiologist Medisch Spectrum Twente, Enschede, The Netherlands.

seizures not respond to medication? Are there perhaps alternatives such as surgery or deep brain stimulation? What triggers her seizures? Can we perhaps develop a device that predicts her seizures?

A 72-year old man has recently been diagnosed with Parkinson's disease. His main complaint is a significant right-sided tremor and problems with walking, in particular stopping and starting is difficult. Sometimes, it is even so severe that he cannot move at all, a phenomenon called freezing. Initially, medication had a fairly good effect on his tremor, with moderate effects on walking. The last years, however, his tremor got worse, walking has almost become impossible, and his symptoms show strong fluctuations during the day. Remarkably, cycling does not pose any problems. What underlies this condition? Can we treat his tremor and walking disability with other means than medication? And what causes the motor symptoms in Parkinson's disease in the first place?

A 23-year old university student was recently diagnosed with a severe mood disorder. Her extremely happy weeks were alternated with depressive periods, and she was eventually diagnosed with a manic-depressive disorder, with mood swings occurring approximately every other two weeks. Treatment with medication had a moderate effect on her mood with several side effects, including blunting of emotion and loss of general interest. We all experience moderate changes in mood, which is normal. In this patient, however, these fluctuations are much stronger. Can we better understand the underlying physiology? Could this understanding contribute to prevention or better treatment? Are there alternatives for drug treatment, for instance, deep brain stimulation?

We discuss neurophysiology and general principles for some of these neurological and neuropsychiatric diseases. Clinically relevant questions vary, but a common element is a change in *dynamics*. Healthy brains switch from normal to abnormal behavior, as in the transition to seizures. What are candidate mechanisms that trigger seizures and why do some patients respond so poorly to current anti-epileptic drugs? Initially, severe injury can suddenly become fatal, as in some patients with stroke. Why do neurons swell in stroke patients and more so in some and hardly in others? Motor behavior can be disturbed by the occurrence of tremors, characterized by involuntary oscillations that are not present in a healthy motor system. How should we treat tremors in patients with Parkinson's disease? Why is deep brain stimulation so effective in some? Moods oscillate between euphoria and depression in patients with a manic-depressive disorder. In other patients, the depressions are so severe that electroconvulsive therapy is the only treatment option left. How does that work?

In the first two chapters, we treat essentials of neurophysiology: the neuron as an excitable cell, action potentials, and synaptic transmission. Next to a treatment of the phenomenology, we present a quantitative mathematical physiological context, including the Hodgkin-Huxley equations. In Chaps. 3 and 4, we introduce scalar and planar differential equations as essential tools to model physiological and pathological behavior of single neurons, This includes a treatment of equilibria, stability, and bifurcations. In this chapter, we also discuss various reductions of the Hodgkin-Huxley equations to two-dimensional models. Chapter 5 describes interacting neurons. We review some fundamental "motifs," treat the integrate-and-fire neuron, and discuss synchronization. In Chap. 6, we introduce the basics of the generation of the EEG, and show various clinical conditions where EEG recordings are relevant. In Chap. 7, we discuss a meanfield model for the EEG, using the physiological and mathematical concepts presented in earlier chapters. Two chapters discuss pathology and include applications of the concepts and mathematical models to clinical problems. Chapter 8 treats dynamics in ischemic stroke including a detailed treatise of processes involved in edema/cell swelling. In Chap. 9, we discuss clinical characteristics of epilepsy, the role of the EEG for diagnostics, and present various mathematical models in use to further understanding of (transition to) seizures. Limitations of current treatment options and pharmacoresistance are treated, as well. Finally, in Chap. 10, we review some clinical applications of neurostimulation. All chapters contain examples and exercises; answers are included.

Mastering the contents of this book provides students with an in-depth understanding of general principles from physiology and dynamics in relation to common neurological disorders. We hope that this enhances understanding of several underlying processes to ultimately contribute to the development of better diagnostics and novel treatments.

Contents

Part I Physiology of Neurons and Synapses

Symbols and Physical Constants

ε_0	Permittivity of free space (8.85 10^{-12}F m^{-1})
ε_r	Relative (static) permittivity ($\varepsilon_c/\varepsilon_0$) (–)
μ	Electrical mobility ($m^2s^{-1}V^{-1}$)
z	Valence (–)
τ	Time constant (s)
r_m	Specific membrane resistance (Ω m^2); Typical value (0.1–1 Ω m^2)
c_m	Specific membrane capacitance (F/m^2); Typical value (0.01 F/m^2)
r_{ax}	Specific axial resistance (Ωm); Typical value (0.1–1 Ωm)
σ	Conductivity (S/m)
R_m	Membrane resistance (Ω)
C_m	Membrane capacitance (F)
D	Diffusion coefficient (m^2 s^{-1})
E	Electrical field strength (V/m)
E_x	Nernst potential of ion x (V)
F	Faraday's constant (96485 C/mol)
g	Conductance (S)
g	Conductance (density) (S/m^2)
I	Current (A)
k	Boltzmann's constant (1.38 10^{-23} J/K)
N	Avogadro's number (6.02 10^{23} molecules/mol)
q	Electron charge (1.6 $\times 10^{-19}$ C)
R	Gas constant (8.31 J/(mol K))
t	time (s)
T	Absolute temperature (Kelvin)
V	Voltage (V)

Part I
Physiology of Neurons and Synapses

Chapter 1
Electrophysiology of the Neuron

The human brain has 100 billion neurons, each neuron connected to 10 thousand other neurons. Sitting on your shoulders is the most complicated object in the known universe.

— Michio Kaku

Abstract In this chapter, we discuss elementary concepts from neurophysiology. We treat the generation of the membrane potential, the role and dynamics of voltage-gated channels and the Hodgkin-Huxley equations. We present experimental techniques, in particular the voltage and patch-clamp technique, that have been essential in elucidating fundamental processes of neuronal dynamics. Several concepts are illustrated with clinical examples. At the end of this chapter, you will understand the critical role of ion concentration gradients for establishing the resting membrane potential and the role of the voltage-gated sodium and potassium channels in the generation of action potentials. You understand how Hodgkin and Huxley were able to formulate the Hodgkin-Huxley equations and you can perform essential simulations using these equations to explore the effects of changes in ion homeostasis or abnormal channel gating.

1.1 Introduction

In this chapter we review key characteristics of neurons, including action potential generation and the role of voltage-sensitive ion channels. This knowledge is essential for an understanding of various pathological conditions, for instance stroke, epilepsy and Parkinson's disease. Recall that neurons, and in fact all human cells, have a trans-

Electronic supplementary material The online version of this chapter (https://doi.org/10.1007/978-3-662-61184-5_1) contains supplementary material, which is available to authorized users.

The original version of this chapter was revised: The belated corrections have been incorporated. The correction to this chapter is available at https://doi.org/10.1007/978-3-662-61184-5_11

M. J. A. M. van Putten, *Dynamics of Neural Networks*,
https://doi.org/10.1007/978-3-662-61184-5_1

membrane potential in the order of −60 to −70 mV, the inside being more negative than the outside. A key characteristic of neurons is that they are excitable. Neurons can quickly change the relative selective permeability of their membrane, causing small ionic fluxes, resulting in fast (1–2 ms) changes in the membrane potential. We will now discuss these processes in more detail. For additional reading, see e.g. [59, 130].

1.2 The Origin of the Membrane Potential

A neuron in rest has a transmembrane voltage of approximately −60 to −70 mV, where the inside is negative with respect to the outside. This transmembrane potential originates from two conditions: transmembrane ion gradients and a semipermeable cell membrane. Four ionic species are most relevant: sodium (Na^+), potassium (K^+), calcium (Ca^{2+}) and chloride (Cl^-). The concentration differences across the cell membrane of all these four ions result in electrochemical gradients. Inside the neuron, the concentration of potassium is relatively large, while the sodium concentration is low. The cell also contains a large concentration of negatively charged macromolecules, P^-. Outside the cell, we find high concentrations of Na^+, Cl^- and Ca^{2+}. This is illustrated in Fig. 1.1. To derive the expression for the membrane potential let us initially assume that the cell membrane is only permeable to K^+ ions, and impermeable to the other three ion species, as illustrated in Fig. 1.2. If the cell membrane potential is initially zero, potassium ions will diffuse from the inside of the cell to the extracellular space, because of the concentration gradient. But as electroneutrality in the bulk needs to be preserved, negative charge accumulates at the inside of the cell membrane, while the potassium ions that leave the cell essentially reside at the outside of the cell membrane. The positive and negative charges are now separated across the cell membrane, thus creating a voltage difference. The K^+ ions now encounter two forces, resulting from (i) the concentration gradient and (ii) the electrical potential gradient.

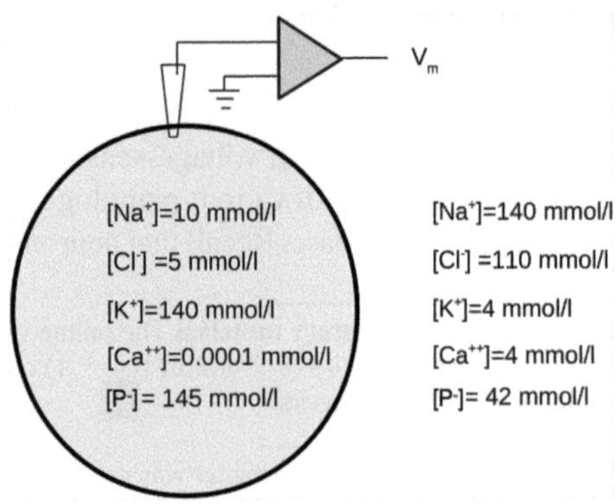

Fig. 1.1 Distribution of major ions across a neuron. Note, that the sum of the (free) positive and negative charges in both the intracellular and extracellular space is zero to conserve electroneutrality. Tiny amounts of charges will be separated across the cell membrane, resulting in a membrane potential V_m, as illustrated in Fig. 1.2

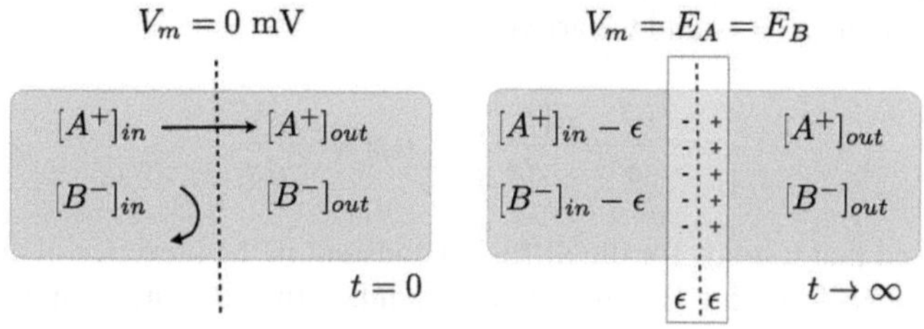

Fig. 1.2 Cartoon of ion gradients and possible fluxes in the calculation of the Nernst potential. The membrane (dashed line) is semipermeable only for the positive ion species A^+ (for instance potassium), and we assume that the intracellular concentration of ion species A^+ is larger than the extracellular concentration. Left: initial condition at $t = 0$; the membrane voltage $V_m = 0$ mV. Right: development of the resting membrane potential. At equilibrium the membrane potential equals the Nernst potentials of the ion species A and B, i.e. $V_m = E_A = E_B$. The charge that accumulates across the membrane, ε, is a tiny fraction of all the charges in the free solution. Note that bulk electroneutrality is preserved, as the same amount of positive and negative charges are now "removed" from the bulk solution, where the positive charges have been able to cross the membrane (as the membrane was semipermeable for these ions only), and now reside on the right side, with negative charges on the left side

Let's add some concreteness to these considerations. We take potassium as our ion of interest and consider a one-dimensional situation, i.e. ions can only move in a single direction, x. For the *diffusion current* of potassium, J_k, across the membrane, it holds that, using Fick's law,

$$J_k(\text{diffusion}) = -D\frac{d[K^+]}{dx} \tag{1.1}$$

where D is the diffusion constant (unit m^2/s). The second force that acts on the potassium ions is an electrical force, resulting from the charge separation across the membrane. This induces a *drift current*, expressed as

$$J_k(\text{drift}) = -\mu z[K^+]\frac{dV}{dx} \tag{1.2}$$

with μ the electrical motility (unit m^{2}s^{-1}V^{-1}), defined as the drift velocity per unit field strength, and z the valence of the ion. For potassium, $z = +1$. For negatively charged ions, z is negative; for example, $z = -1$ for Cl$^-$. For the total potassium current it now holds that

$$J_{\text{K,tot}} = -D\frac{d[K^+]}{dx} - \mu z[K^+]\frac{dV}{dx}. \tag{1.3}$$

In equilibrium, $J_{\mathrm{K,tot}} = 0$, and we arrive at

$$-\frac{kT}{q}\mu\frac{d[K^+]}{dx} = \mu z[K^+]\frac{dV}{dx} \tag{1.4}$$

where we used that it holds for the diffusion constant in the presence of an electrical field that $D = \mu\frac{kT}{q}$, with k Boltzmann's constant, T the absolute temperature and q the elementary charge (Einstein relationship). Setting $z = 1$, we write

$$\frac{dV}{dx} = -\frac{kT}{q[K^+]}\frac{d[K^+]}{dx}. \tag{1.5}$$

Integrating across the membrane from a place x_o on the outside to a place x_i on the inside, we have

$$\int_{V_o}^{V_i} dV = -\frac{kT}{q}\int_{[K^+]_o}^{[K^+]_i}\frac{d[K^+]}{[K^+]} \tag{1.6}$$

which results in the equilibrium potential

$$\Delta V = V_i - V_o = E_{\mathrm{Nernst}} = \frac{kT}{q}\ln\frac{[K^+]_{\mathrm{out}}}{[K+]_{\mathrm{in}}}\ \mathrm{V} \tag{1.7}$$

with $[K^+]_{\mathrm{in}}$ and $[K^+]_{\mathrm{out}}$ the ion concentration of potassium inside and outside of the cell, respectively. Note, that we define the transmembrane potential difference as the intracellular voltage minus the extracellular voltage. This expression is known as the Nernst equation and the resulting value as the Nernst potential. In other texts, you may read that $E_{\mathrm{Nernst}} = \frac{RT}{F}\ln\frac{[K^+]_{\mathrm{out}}}{[K+]_{\mathrm{in}}}$, which is identical as $k/q = R/F$. If we set the temperature to 37 °C ($T = 310$ K), and substitute the numbers for the other constants, we find that

$$E_{\mathrm{Nernst}} \approx 27 \cdot \ln\frac{[K^+]_{\mathrm{out}}}{[K^+]_{\mathrm{in}}}\ \mathrm{mV} \tag{1.8}$$

or

$$E_{\mathrm{Nernst}} \approx 62 \cdot \log_{10}\frac{[K^+]_{\mathrm{out}}}{[K^+]_{\mathrm{in}}}\ \mathrm{mV}. \tag{1.9}$$

For potassium, the Nernst potential is approximately −96 mV, using an intracellular concentration of 140 mmol/l and an extracellular concentration of 4 mmol/l. The Nernst potentials for the different ion species are typically written as E_{Na}, E_{K}, E_{Cl}, and E_{Ca} for the sodium, potassium, chloride and calcium ions, respectively. The Nernst potential is an *equilibrium potential*: at this transmembrane potential difference there is a *zero net flux for the permeable ion*. The Nernst potential, therefore, gives the potential arising when a single permeant species reaches equilibrium. For more details, see e.g. [59, 130]. This is a good moment to make Exercise 1.1.

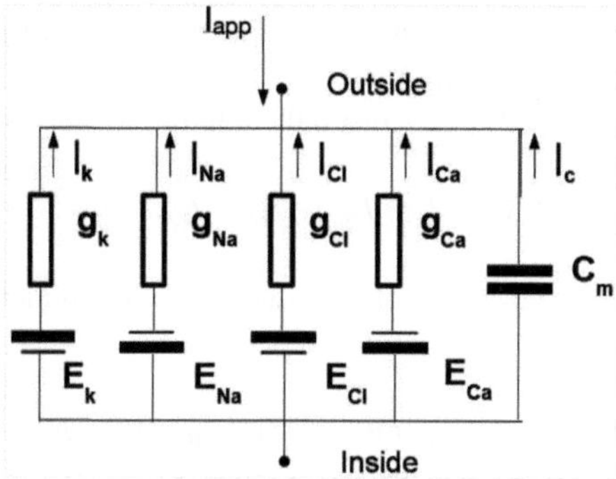

Fig. 1.3 Electrical equivalent circuit of a part of the cell membrane, showing four batteries, representing the Nernst potentials of potassium, sodium, chloride and calcium. The membrane capacitance is represented by the capacitor C_m. A potential additional current I_{app}, e.g. from synapses, is indicated, too. The sodium-potassium pump (to be discussed later) and the leak current are not shown. Recall that conductance (unit S) is the reciprocal of resistance (unit Ω)

1.2.1 *Multiple Permeable Ions*

In the real world, the cell membrane is not permeable for a single ion species and impermeable for the other ions, but shows differential permeabilities (none equal to zero) of the various ions. This results in a transmembrane potential that is a *weighted sum of the different Nernst potentials*. The weighting factor depends on the relative conductances[1] of the various ions.

We introduce the electrical equivalent circuit of the cell membrane, Fig. 1.3. Shown are four batteries that represent the different Nernst potentials, each with a series resistor with conductance, g. The cell membrane capacitance is represented by a capacitor, C_m. The resting membrane potential (equilibrium potential) is a weighted sum of the different Nernst potentials, given by

$$V_{\text{rest}} = \frac{g_{\text{Na}} E_{\text{Na}} + g_{\text{K}} E_{\text{K}} + g_{\text{Ca}} E_{\text{Ca}} + g_{\text{Cl}} E_{\text{Cl}}}{g_{\text{Na}} + g_{\text{K}} + g_{\text{Ca}} + g_{\text{Cl}}}. \tag{1.10}$$

We can now also derive an expression for the membrane currents, using Kirchoff's law, that results in

$$C\dot{V} = I - \sum I_{\text{ion}} \tag{1.11}$$

[1] Here, we take the approach using conductances after introducing the electrical equivalent circuit. A similar approach is to use the Goldman-Hodgkin-Katz voltage equation that is based on ion concentrations and permeabilities. See e.g. [130].

with I an additional current (indicated with I_{app} in Fig. 1.3). Recall that the capacitative current is given by the membrane capacity, C, multiplied by the change in the membrane voltage $\frac{dV}{dt} = \dot{V}$. We now insert the four main ionic currents in the previous equation

$$C\dot{V} = I - I_{\text{Na}} - I_{\text{K}} - I_{\text{Ca}} - I_{\text{Cl}} \tag{1.12}$$

and by writing each ionic current as the product of their conductance and the voltage difference, we obtain

$$C\dot{V} = I - g_{\text{Na}}(V - E_{\text{Na}}) - g_{\text{K}}(V - E_{\text{K}}) - g_{\text{Ca}}(V - E_{\text{Ca}}) - g_{\text{Cl}}(V - E_{\text{Cl}}). \tag{1.13}$$

If we now set the external current, $I = 0$, it is straightforward to derive (1.10). Check this yourself!

As an alternative expression for the change in membrane voltage as a function of the external current, I, and the various conductances, we can also write

$$C\dot{V} = I - g_{\text{input}}(V - V_{\text{rest}}), \tag{1.14}$$

where

$$g_{\text{input}} = g_{\text{Na}} + g_{\text{K}} + g_{\text{Cl}} + g_{\text{Ca}} \tag{1.15}$$

is the total conductance or *input conductance*. Prove this yourself in Exercise 1.2.

It follows from (1.14) that the membrane voltage V continuously *tends to go* to the value given by

$$V \to V_{\text{rest}} + IR_{\text{input}} \tag{1.16}$$

that is obtained by setting $\dot{V} = 0$. Note, that we used the reciprocal value of the conductances, the membrane resistance $R_{\text{input}} = 1/g_{\text{input}}$. Here, $V_{\text{rest}} + IR_{\text{input}}$ is a *globally attracting equilibrium*, that will be discussed in more detail in Chap. 4. The interpretation is that an *external current changes the membrane voltage*, where the amount of change from the initial steady state is defined by the product of this current and the input resistance of the membrane (Ohm's law). If the conductances would be constant, as perhaps naively suggested by (1.13), there would be no interesting dynamics: we would only observe exponential convergence to an equilibrium, only depending on the input current. However, conductances are not constant, but depend on the membrane voltage. This is the origin of the possibility for nontrivial dynamics. It makes the system (1.13) a truly *nonlinear system*, allowing for instance the generation of spikes, as we will discuss later in this chapter and the Chaps. 3 and 4.

1.2.2 Active Transport of Ions by Pumps

Remember that the rest membrane potential, given by (1.10), is typically not equal to the various Nernst potentials of the ion species involved. This implies that, even if the neuron is 'at rest', the individual ion currents are not equal to zero. If we take potassium as an example, it holds that

$$I_k = g_k(V_{\text{rest}} - E_{\text{K}}) \neq 0 \tag{1.17}$$

because $V_{\text{rest}} \neq E_{\text{K}}$ and the membrane conductance, g_{K}, has a nonzero value at this membrane potential. This holds for all other ions involved, mainly sodium and chloride, too. Therefore, ions constantly move through the membrane by diffusion, which would result in a drift in the membrane potential and loss of ion gradients. This is not observed, however: these passive currents are compensated by active transport of ions by pumps, that obtain energy from the hydrolysis of ATP.[2]

1.2.3 ATP-Dependent Pumps

Cells have many ATP dependent pumps to maintain and restore ion gradients. The sodium-potassium pump actively pumps three Na^+ ions outwards for each two K^+ ions, that are carried into the cell. In addition to the Na/K-ATPase pump, cells have various other energy dependent pumps, for instance the Ca-ATPase, that maintains the very low intracellular Ca^{2+} concentration (about 10.000 times lower than the extracellular concentration, except for short moments just preceding the excretion of a neurotransmitter, discussed in detail in Chap. 2). Other transport mechanisms in the cell membrane are *transporters*, that are not ATP dependent. These transporters carry Na^+, K^+ or Cl^- ions along their electrochemical gradients, and the energy of these gradients can be used to *antiport* other ions against their concentration gradient. Examples are the Na-Ca transporters (Na^+ is carried inwards (*symport*), and Ca^{2+} outwards, (*antiport*)) and the K-Cl transporter KCC, transporting K^+ outwards and Cl^- outwards. As we will learn in later chapters, energy depletion will affect these pumps, with significant impact on neuronal function as membrane potentials will not be maintained. This is observed in various clinical conditions, for instance in patients with a stroke, cardiac arrest, metabolic encephalopathies and primary mitochondrial diseases.

[2]The pump also has a small contribution to the membrane potential, as the net current is not zero (three sodium ions are pumped out while two potassium ions are pumped in). The effect is small, however, approximately −2 to −5 mV.

1.3 Neurons are Excitable Cells

An important property of the cell membrane of neurons is that the ionic conductances can change in response to various input stimuli, which makes neurons *excitable cells*: neurons can quickly change the voltage across (a local part) of the cell membrane, which results from changes in the *semipermeable characteristics* of the cell membrane. Ion channels consist of large proteins, where the opening (the *pores*) is controlled by various *gates*, that switch the channels between an open or a closed state. In this way, the sodium conductance, g_{Na}, can change from 0 to 120 mS/cm^2 and the potassium conductance g_K from 0 to 40 mS/cm^2. The channel gates are controlled by various means. The three main categories are i) the *voltage-gated* channels (e.g. voltage-gated Na^+ or K^+ channels); ii) *ligand-gated* channels e.g. the GABA or the acetylcholine receptor, that allows Cl^- and (mainly) Na^+ ions to flow along their electrochemical gradients, respectively, and iii) channels that are controlled by second messengers. An example of the latter category is a Ca^{2+}-gated potassium channel.

1.3.1 Voltage-Gated Channels

The conductance g of the various voltage-gated channels depends on the membrane potential. The net current generated by a large population of identical channels[3] can be described by

$$I = \overline{g}p(V - E) \tag{1.18}$$

with E the Nernst or *reversal* potential of the particular ion, and $g = \overline{g}p$ with $\overline{g}$ the *maximal conductance* of the population and $p \in [0, 1]$ the voltage-dependent proportion of open channels in the population. The voltage dependence of the ion conductances is illustrated in Fig. 1.4 for potassium and sodium. It is shown that as the voltage increases, the potassium conductance increases, resulting in a *persistent current*. The sodium conductance shows a different behavior: it initially increases, but subsequently, despite the persistent increased membrane potential, the conductance returns to baseline. Thus, Na^+ channels first activate and then inactivate, generating a *transient current*. These characteristics are essential for the generation of the action potential, discussed next.

[3] We are dealing with average properties. At a later instance, we discuss the behaviour of individual ion channels.

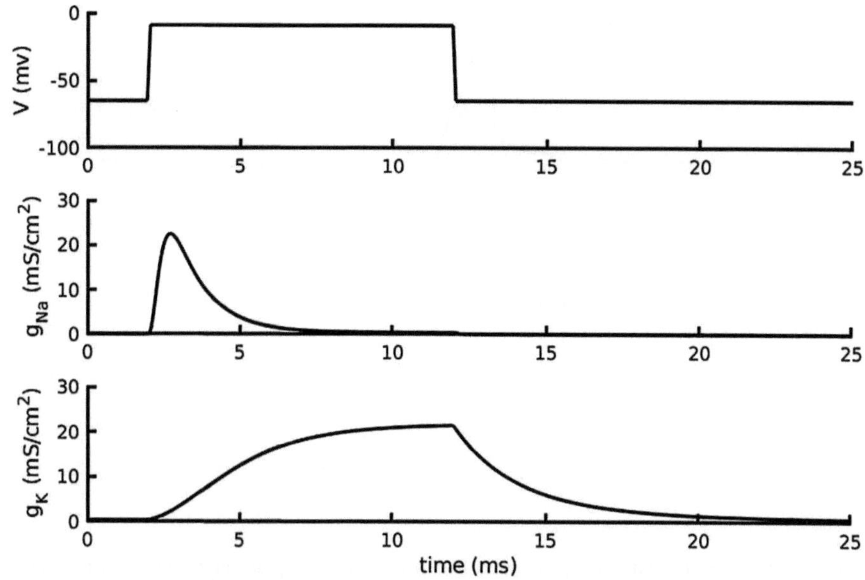

Fig. 1.4 Shown are the sodium and potassium conductances, g_{Na} and g_K, respectively (middle and lower panel), as a function of time after a change in the membrane potential from −65 mV to −9 mV (top panel) as could be recorded in the squid axon (simulated data). Note that the sodium conductance, g_{Na}, after an initial increase, returns to a value near zero, while the potassium conductance, g_K, remains increased. While the graph may suggest that both the sodium and potassium conductances are zero at baseline, this is not the case; $g_k \approx 0.4$ mS/cm^2 and $g_{Na} \approx 0.01$ mS/cm^2. In this particular experiment, the voltage clamp technique was used that can set the membrane potential to a predefined value, in this case −9 mV. Also note the differences in time constants for the rise and decay of the channels. The voltage clamp technique, used to obtain these experimental results, is discussed in Sect. 1.4

1.3.2 The Action Potential

Action potentials are generated by the dynamical interplay between sodium and potassium fluxes, resulting from changes in their conductances, that are controlled by the voltage-gated Na^+ and K^+ ion channels. In Fig. 1.5 the time-course of an action potential is shown, with the associated potassium and sodium conductances, g_K and g_{Na}. It was the combined effort of Hodgkin and Huxley that resulted in an explicit description of the equations that describes the dynamics of the conductances of these voltage-gated channels. These equations, published in 1952, are known as the Hodgkin-Huxley equations. Hodgkin and Huxley discovered that as the membrane voltage changes, the permeability for sodium and potassium ions changes, as well: there is a voltage dependency of the various ionic conductances, as illustrated in Fig. 1.4. They used very clever experimental techniques to study the time-dependent changes in membrane voltage of the squid giant axon. The control variables introduced in their equations are essentially *gates* that can open or close, depending on the membrane voltage.

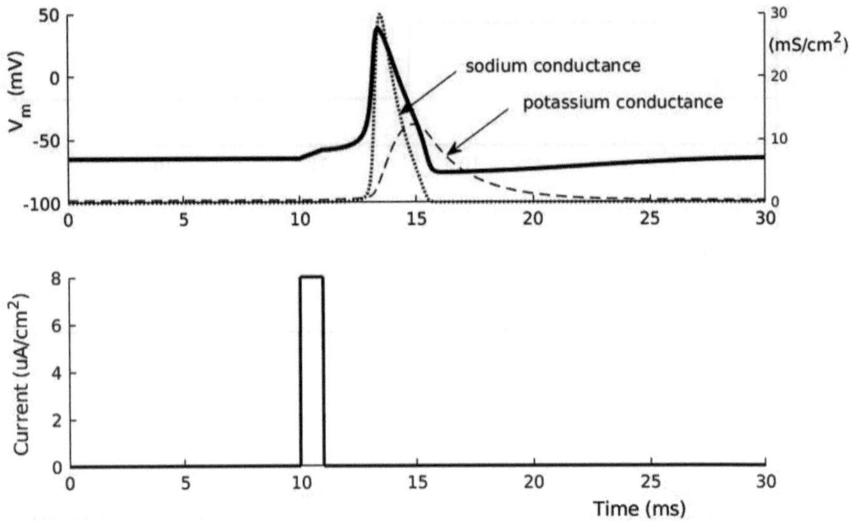

Fig. 1.5 Time course of an action potential (top panel, solid curve) generated in response to a membrane current pulse (bottom), showing the conductances of the two major currents underlying its generation. Transient sodium currents cause the depolarization, while subsequently both the transient nature of the sodium current and the increased potassium conductance caused by the depolarization result in a relatively fast restoration of the membrane potential. As the potassium conductance remains temporarily increased due to its slower time constant (cf. Fig. 1.4), the membrane potential is temporarily even lower than the resting membrane potential

The dependency on the voltage appeared to define both the final position of the gate (is it completely open or closed or only partially) and the velocity of the gate to reach this final position. This is illustrated in Fig. 1.6. In the case of a single gate, the channels open or close as a function of a control voltage. For the voltage gated potassium channels, the gate is labeled with the activation variable $n \in [0, 1]$. For the transient sodium current, two gates are needed: one gate opens in response to an increase in voltage (activation variable, m) while the other gate will close in response to an increase in voltage (inactivation variable, h), with $m, h \in [0, 1]$. The speed of gate opening of closing is further described by a time constant, τ, that is also voltage dependent. Each gate has its own time constant. For the sodium channel, the time constant of the 'm-gate' is smaller than that of the inactivation variable, h. Clearly, it would be pointless if the h-gate would have the smallest time-constant: opening of the channel should be faster than closing it, otherwise the channel could not conduct a current (cf. Fig. 1.6). We will now discuss the explicit voltage dependency of the gating variables and time constants to finally arrive at the Hodgkin Huxley equations.

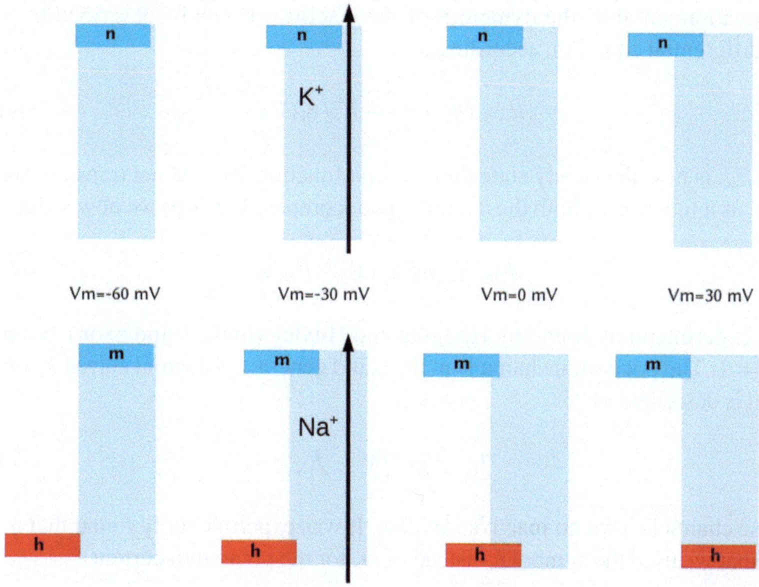

Fig. 1.6 Cartoon of the persistent potassium and transient sodium ion channel. The actual channel that allows ion passage is sketched as light blue. Top: potassium ion channel with an activation variable n, only. Bottom: ion channel for the transient sodium current, with both an activation and an inactivation variable, m and h as a function of typical membrane voltages. Depending in the voltage, the gate opens or closes to a final position, expressed as a value between fully closed (0) and completely open (1). The top row illustrates that the voltage-gated K^+ channel is slightly open at a membrane potential of −60 mV, while the voltage-gated sodium channel is essentially closed at −60 mV

1.3.3 *Quantitative Dynamics of the Activation and Inactivation Variables*

The activation variable for the sodium channel, m, is described by a first-order non-linear differential equation[4]

$$\dot{m} = (m_\infty(V) - m)/\tau_m(V) \tag{1.19}$$

with time constant $\tau_m(V)$ and steady state activation $m_\infty(V)$. We use the notation $\dot{m}$ to express the time derivative of m that can also be written as $\frac{dm}{dt}$. The rate of change in the variable m in response to a change in membrane voltage is defined by the time constant, while the asymptotic value of m equals m_∞. To make matters a little more complicated, the time constant is a function of the voltage, too. Both the asymptotic value m_∞ and the time constant can be measured by voltage-clamp experiment, discussed later in this chapter. Similar to the dynamics of

[4]We discuss general aspects of differential equations in Chap. 3.

the activation variable, the dynamics of the inactivation can be described by a first order differential equation, according to

$$\dot{h} = (h_\infty(V) - h)/\tau_h(V), \tag{1.20}$$

where h_∞ is now the steady state *inactivation* function. For the net transient sodium current as a function of both the m and h gate (compare Fig. 1.6) we now write

$$I_{\mathrm{Na}} = \overline{g} m^a h^b (V - E_{\mathrm{Na}}). \tag{1.21}$$

It was experimentally found by Hodgkin and Huxley (in the squid axon) that $a = 3$ and $b = 1$. The potassium channels in the squid generate persistent currents, and this current is described as

$$I_{\mathrm{K}} = \overline{g} n^a (V - E_{\mathrm{K}}) \tag{1.22}$$

as these channels have no inactivation gate. It was experimentally found that $a = 4$. Note, that we used the symbol n instead of m, for the potassium current.

1.3.4 The Hodgkin-Huxley Equations

For the squid axon that has sodium and potassium channels, only, and using (1.13) (without calcium and chloride channels), we now present the Hodgkin-Huxley equations [52]:

$$C\dot{V} = I - \overline{g}_{\mathrm{K}} n^4 (V - E_{\mathrm{K}}) - \overline{g}_{\mathrm{Na}} m^3 h (V - E_{\mathrm{Na}}) - g_L (V - E_L), \tag{1.23}$$

where we added an Ohmic (g_L = constant) leak current, with reversal potential E_L, and where the activation and inactivation variables are given by[5]

$$\begin{aligned} \dot{n} &= (n_\infty(V) - n)/\tau_n(V), \\ \dot{m} &= (m_\infty(V) - m)/\tau_m(V), \\ \dot{h} &= (h_\infty(V) - h)/\tau_h(V) \end{aligned} \tag{1.24}$$

and steady state values and corresponding time constants given by

$$\begin{aligned} n_\infty &= \alpha_n/(\alpha_n + \beta_n) \ , \ \tau_n = 1/(\alpha_n + \beta_n), \\ m_\infty &= \alpha_m/(\alpha_m + \beta_m) \ , \ \tau_m = 1/(\alpha_m + \beta_m), \\ h_\infty &= \alpha_h/(\alpha_h + \beta_h) \ , \ \tau_h = 1/(\alpha_h + \beta_h) \end{aligned} \tag{1.25}$$

[5]This is the standard form of the activation and inactivation variables.

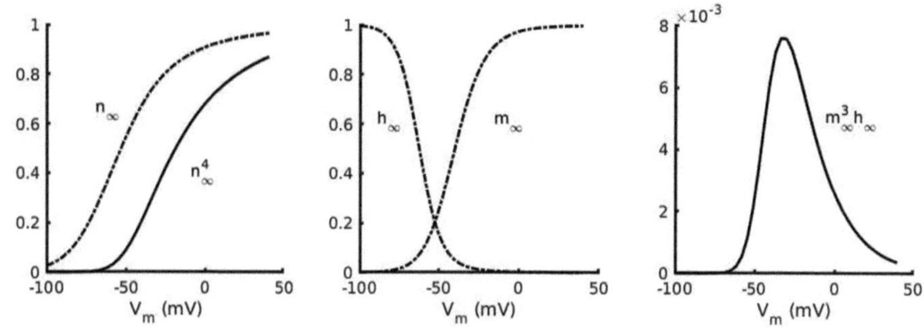

Fig. 1.7 Left panel: Steady-state activation function n_∞ and n_∞^4 for potassium as a function of membrane voltage. Middle panel: activation function m_∞ and inactivation h_∞. Right panel:the value of $m_\infty^3 h_\infty$. Note, that the sodium conductance is practically zero at V_{rest}. The membrane potential has been shifted to its true value (i.e. the voltage in the extracellular space is set to 0), so that the resting membrane potential is near −65 mV

with

$$\begin{aligned} \alpha_n &= 0.01(V+55)/(1-\exp(-(V+55)/10)), \\ \beta_n &= 0.125\exp(-(V+65)/80), \\ \alpha_m &= 0.1(V+40)/(1-\exp(-0.1(V+40))), \\ \beta_m &= 4\exp(-(V+65)/18), \\ \alpha_h &= 0.07\exp(-(V+65)/20), \\ \beta_h &= 1/(1+\exp(-0.1(V+35))). \end{aligned} \tag{1.26}$$

Note, that these equations are not the same as in the original papers of Hodgkin and Huxley; they described the *change* of the voltage from the resting potential (set to zero). The equations presented here will result in a resting membrane potential $V = -65\,\text{mV}$ using $g_{\text{Na}} = 120\,\text{mS/cm}^2$, $g_{\text{K}} = 36\,\text{mS/cm}^2$, $g_{\text{L}} = 0.3\,\text{mS/cm}^2$, $E_{\text{Na}} = 50\,\text{mV}$, $E_{\text{K}} = -77\,\text{mV}$ and $E_{\text{L}} = -54.4\,\text{mV}$. Note further that we dropped the explicit voltage dependence of the steady state values and time constants in the equations. The *steady state* activation and inactivation functions are shown in Fig. 1.7. The voltage-dependent time constants of the Hodgkin-Huxley model, defined in (1.25), are illustrated in Fig. 1.8. Note that the time constant for the m-gate is small (and therefore the channel opening is fast), while the n and h gating is relatively slower.[6] A simulation of the action potential with the dynamics of the various gating variables is shown in Fig. 1.9. In response to the depolarization, induced by the external current, the activation variables m and n increase and h is decreased. But because the time constant

---

[6]Historically, Hodgkin and Huxley used different expressions for the activation and inactivation variables, using

$$\begin{aligned} \dot{n} &= \alpha_n(V)(1-n) - \beta_n(V)n, \\ \dot{m} &= \alpha_m(V)(1-m) - \beta_m(V)m, \\ \dot{h} &= \alpha_h(V)(1-h) - \beta_h(V)h, \end{aligned}$$

where the functions $\alpha_j(V)$ and $\beta_j(V)$, with $j \in n, m, h$ describe the transition rates between open and closed states of the channels.

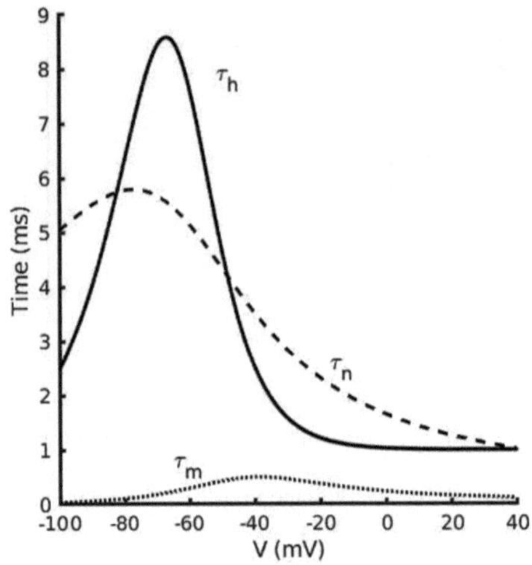

Fig. 1.8 Voltage-dependent time constants of the Hodgkin-Huxley model. The membrane potential has been shifted to its true value, so that the resting state is near -65 mV. Note that the time constant for the sodium activation variable, τ_m, is much smaller than the activation variable for the potassium, τ_n and the inactivation of the sodium channel, τ_h. This results in a fast opening of the Na^+ channel. The subsequent closing of the Na^+channel and opening of the K^+channel is relatively slower

for the m-gate, τ_m, is small, the m gate responds relatively fast to the depolarization, resulting in an activation of the sodium conductance, that on its turn further activates g_{Na}. This positive feedback loop finally results in an increase of the membrane voltage, and the upstroke of the action potential. Now, however, the slower gating variables start to change their values, resulting in an inactivation of the sodium current as h drops to 0 and an activation of the potassium current as n increases towards 1. This results in a repolarization towards the resting membrane voltage. Because the time constant for the gating variable n for the potassium current is slow, the membrane voltage will temporarily be below its resting value, approximating the Nernst potential of potassium. This phenomenon is known as *afterhyperpolarization*. During this period, it is not possible to generate another action potential: this is the *absolute refractory* period. The interval thereafter is referred to as the *relative refractory* period, where action potentials can be generated if the stimulus is strong enough.

1.4 Voltage Clamp

We did not discuss yet how Alan Hodgkin and Andrew Huxley arrived at their famous equations. They worked together in 1939, and again from 1946 to 1952, and are known as one of the most productive and influential collaborators in the history of physiology. Hodgkin and Huxley chose the giant squid axon as a model system for their

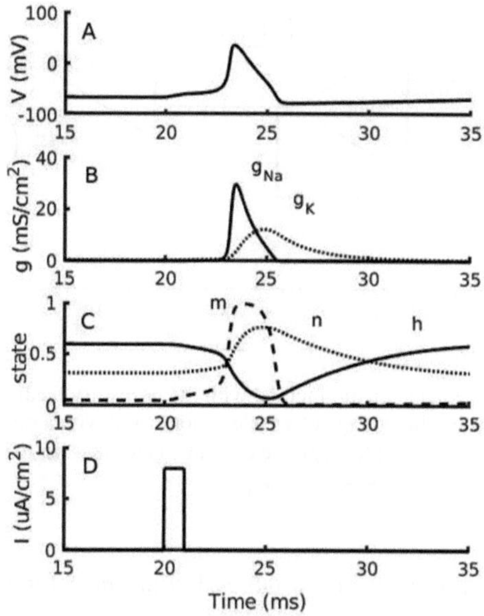

Fig. 1.9 Simulation of the generation of an action potential (**a**) in response to an excitatory current (**d**), with the dynamics of conductances (**b**) and the gating variables (**c**). Compare the time course of the gating variables m, h, n with the voltage dependency shown in Fig. 1.8, noting that the m-gate is indeed relatively fast. Note also, that the membrane potential at the end of the action potential is more negative than at the start, resulting from the increase in potassium current (the n-gate is still relatively open), and the membrane voltage tends towards the Nernst potential of potassium

experiments. Its large size with a diameter up to 0.5 mm allowed the insertion of small wire inside the axon (essential for the voltage clamp measurements), and the ability to survive for many hours in a laboratory environment. Also, as it turned out later, the number of different ion channels in the squid axon is very limited. Indeed, it only contains sodium and potassium channels, making it an ideal structure to study the voltage-dependent permeability of its cell membrane. Experimentally, it was a tremendous challenge to discover the various voltage-dependent components in these equations, responsible for the generation of the action potential. However, the voltage clamp method, a technique that had been devised in the 1930s by Cole and Curtis, allowed major breakthroughs in isolating these various voltage-dependencies: the voltage-clamp allows the membrane voltage to be set to a predefined, fixed value, *even if there are changes in the conductances*. This is realized by compensating any change in ionic current due to changes in transmembrane conductance by the injection of another current. The current injected into the axon is then the mirror image of the current generated by the changes in conductivity at that potential (Fig. 1.10). Why is this technique essential to understand better what controls the various ionic currents? Hodgkin and Huxley were aware that the ionic currents are both time- and voltage-dependent. To separate the time- and voltage-dependency performing experiments by which the voltage could be controlled allowed them to describe the voltage dependent

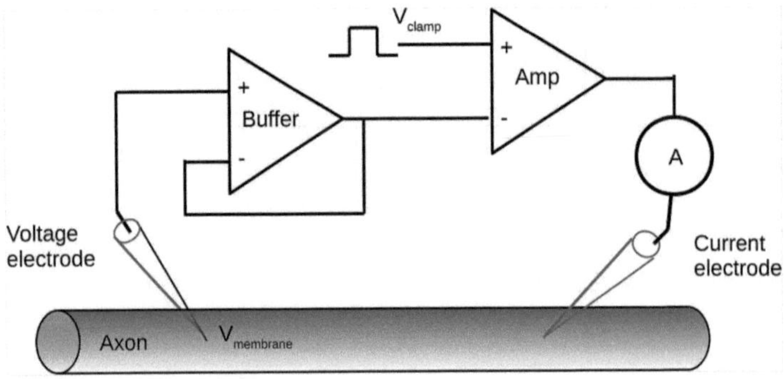

Fig. 1.10 Cartoon of the voltage clamp method with the necessary feedback circuitry. The membrane voltage measured with the voltage electrode is compared with a reference value (the clamp voltage, V_c and the current injected through the current electrode is set to such a value, controlled by the feedback circuitry, that $V_{\text{membrane}} = V_{\text{clamp}}$

characteristics of the various gating variables. For instance, for the potassium current they discovered that it can be described as

$$I_K(V,t) = g_{\text{K}}(V(t) - E_K) = \overline{g}_{\text{K}} n^4 (V(t) - E_K). \tag{1.27}$$

Note that we have explicitly indicated the voltage- and time-dependence of the potassium current I_K. This equation states that the potassium current is a function of the difference in membrane voltage, V, and the potassium reversal potential, E_K, multiplied by the maximum conductance, $\overline{g}_{\text{K}}$ and the gating variable, n, where n satisfies (cf. (1.25))

$$\dot{n} = \frac{n_\infty(V) - n}{\tau_n(V)}. \tag{1.28}$$

The gating variable controls the time course of the opening and closing of the potassium channel, and is a function of the membrane voltage, as well. This is why we explicitly indicated the voltage dependence of n_∞ and the time constant, τ_n in (1.28). Suppose now, that we would like to find the values of n_∞ and τ_n: this is complicated because both n_∞ and τ_n are a function of the membrane voltage. If we could keep the voltage constant, however, the solution of the ordinary differential equation (1.28) is

$$n(t) = n_\infty(V) - (n_\infty(V) - n(0))e^{-t/\tau_n(V)}. \tag{1.29}$$

Using the voltage-clamp technique, Hodgkin and Huxley measured the current as function of a defined voltage step. Waiting sufficiently long allows a stationary situation to be reached where I_K=constant. In that case, it holds that

$$I_K = \overline{g}_K n_\infty^4 (V - E_K) \tag{1.30}$$

since it follows directly from (1.29) that $\lim_{t\to\infty} n(t) = n_\infty(V)$. But wait! How did Hodgkin and Huxley know in their experimental setup that they were only dealing with a potassium conductance? In the squid axon, containing both potassium and sodium channels, application of different voltage steps in the voltage clamp measurements will not only influence the voltage dependent potassium channels, but also the voltage dependent sodium channels. Here, additional smart experimentation came into play. It is possible, for instance, to *block* all sodium channels with tetrodotoxin (TTX). By performing *two* experiments, one without TTX and one with TTX, results in two voltage dependent currents. The current without TTX is the sum current of the sodium and potassium ions; the current with TTX is the potassium current, only. Subtraction of the two currents then results in the voltage dependent sodium current, from which, of course, the potassium current can subsequently be estimated, as well. Another possibility is to remove a percentage of the extracellular sodium, creating a sodium Nernst potential equal to the clamp voltage. The membrane current measured at this value of the voltage clamp will only contain a potassium (and a small leak) current, as at this membrane voltage the sodium current vanishes.[7] By, again, combining measurements with the physiological sodium concentration (100%) and the concentration where its Nernst potential is equal to the clamp voltage, currents can be separated, as illustrated in Fig. 1.11.

1.5 Patch Clamp

The voltage clamp instrumentation used by Hodgkin and Huxley resolved *aggregate* currents, that were the result of currents flowing through many channels (thousands). In biological reality, ion currents flow through individual channels. This was discovered by a refinement of the technique used by Hodgkin and Huxley. By using a very small pipette that measures the ionic currents through a tiny part of the cell membrane, only, it became possible to measure ion currents through single channels. This technique was invented by Neher and Sakmann, who shared the Nobel Prize in Physiology or Medicine in 1991 for their work on 'the function of single ion channels in cells' and invention of the *patch clamp*. The patch clamp is basically a refinement of the voltage clamp technique. Using this technique, it is possible to record currents of *single ion channels*. The electrode used is a glass micro-pipette that has an open tip with a diameter of about one μm, a size enclosing a membrane surface area or 'patch' that often contains just one or a few ion channel molecules. An illustration of a cell patch is shown in Fig. 1.12. This approach had a major impact on the study of membrane

[7]Remember that it then holds that $I_{\text{Na}} = g_{\text{Na}}(E_{\text{Na}} - V_m) = g_{\text{Na}} \times 0$ as the Nernst potential of sodium is experimentally set to equal the membrane voltage set by the voltage clamp.

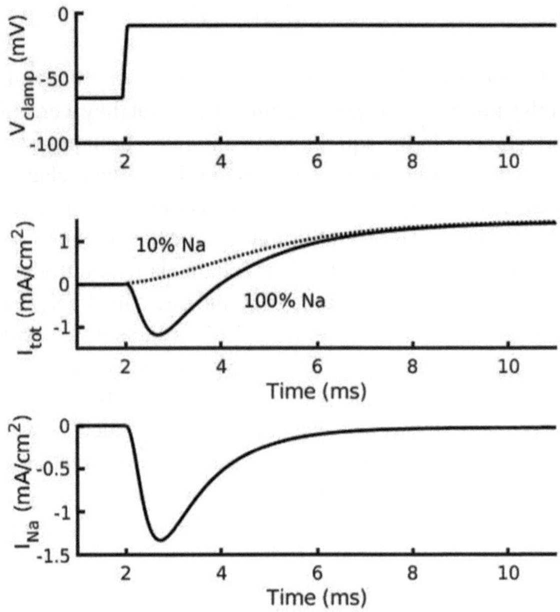

Fig. 1.11 Upper panel: voltage clamp set to -9 mV at t $= 2$ ms. Middle panel: time course of transmembrane currents as a function of two concentrations of sodium. The physiological concentration (100%) is equal to a Nernst potential of approximately 50 mV. The solid line is the sum current of sodium and potassium. The 10% condition has an associated sodium Nernst potential of -9 mV, the same value as the clamp voltage, and the dotted line represents the potassium current, only. Lower panel: time course of the sodium current obtained by subtracting the two currents displayed in the middle panel

currents, and allowed the first *direct* evidence for the presence of voltage-dependent ion-selective channels. The size of individual currents is minuscule, a few pA, only. A historical recording of these microscopic currents, measured with the patch clamp technique, is shown in Fig. 1.13.

1.5.1 Relation Between Single Ion Channel Currents and Macroscopic Currents

The currents that flow through the individual ion channels are all-or-nothing currents of small amplitude, of the order of pico Amperes (10^{-12} A). Macroscopic currents, as recorded by Hodgkin and Huxley, result from many (thousands or more) ion channels, where the average statistical voltage-dependent properties of each ion channel are described by the activation and inactivation variables n, m and h we introduced ear-

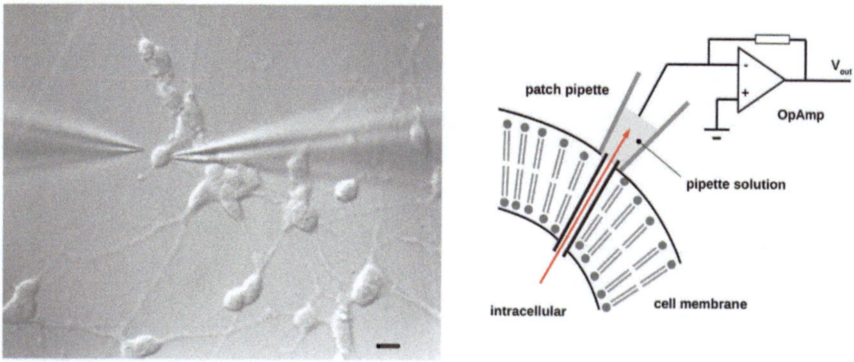

Fig. 1.12 Left: A phase contrast image of a cultured mouse hippocampal neuron which is patch-clamped in the whole cell mode (pipette from the right). The second pipette (from the right) is a stimulation pipette for focal drug application, e.g. glutamate. Horizontal bar is 10 μm. Courtesy of Dr. Karl Kafitz, Institute of Neurobiology, Heinrich-Heine-Universität, Düsseldorf, Germany. Right: Cartoon of the patch-clamp technique. The pipette encloses a single channel, allowing the measurement of a single channel current

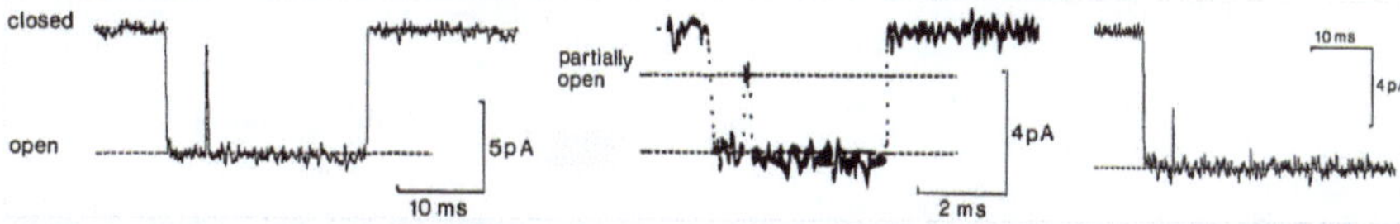

Fig. 1.13 High-resolution current recording of single-channel currents activated by low concentrations ($\ll$500 nM) of acetylcholine at the neuromuscular endplate of frog muscle fibers. Left panel: The open channel current is interrupted by a brief closing gap, which is followed by reopening of the channel (Nachschlag). Middle panel: The opening is followed by a brief current step towards a substate of conductance where the channel is only partially open (conductance substate). Right panel: While the channel is open, the trace shows increased "noisiness" (open channel current noise). Slightly modified illustration and caption from: Sakmann and Neher, Patch clamp techniques for studying ionic channels in excitable membranes, Ann Rev of Physiology, 1984. Reprinted with permission from the American Physiological Society

lier in (1.24)–(1.26). While Hodgkin and Huxley suggested that ions may flow through particular channels, the individual ion currents were only measured after the introduction of the patch-clamp technique by Sakmann and Neher. The relation between the single ion channel currents and the macroscopic current for the persistent K^+-ion channel in a squid axon is illustrated in Fig. 1.14. In this particular experiment, the sodium channels were blocked by tetrodotoxin. Note, that channels open with a variable delay, and most remain open as long as the membrane potential is maintained at +50 mV. This is contrasted with ionic currents through the Na^+ channel, as illustrated in Fig. 1.15. In this experiment, the potassium channels were blocked, and all currents now result from sodium channel opening and closing. While most channels open shortly after the membrane is depolarized to approximately −10 mV, the probability

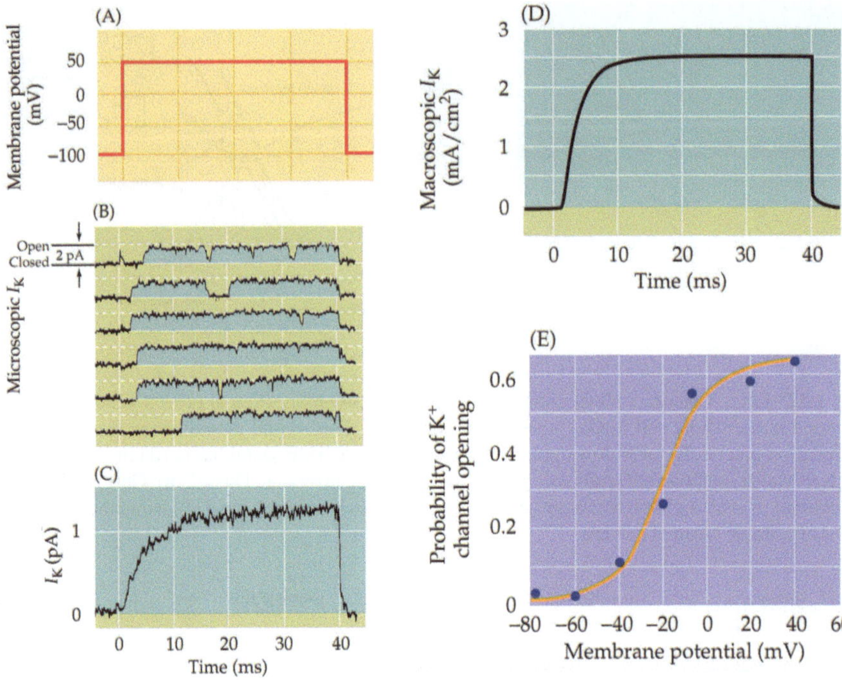

Fig. 1.14 Single ion currents and macroscopic current for a persistent channel current. A. membrane voltage applied in a voltage-clamp experiment. B. individual ion currents. C. Mean ion channel current. D. Total current. E. Probability of channel opening as a function of the membrane voltage, described by the steady state n_∞ activation function in the Hodgkin-Huxley equations. Illustration from Neuroscience, 3rd edition (editors Purves et al.). Reproduced with permission from Oxford Publishing Limited

of remaining open decreases and eventually all channels inactivate, characteristic of a transient current.

1.6 Summary

In this chapter we discussed the generation of the membrane potential and the Hodgkin-Huxley formalism. We treated the electrical equivalent circuits for the excitable cell membrane. Voltage- and patch-clamp techniques made essential contributions to our understanding of the voltage dependent permeabilities of neuronal membranes and ion channel function.

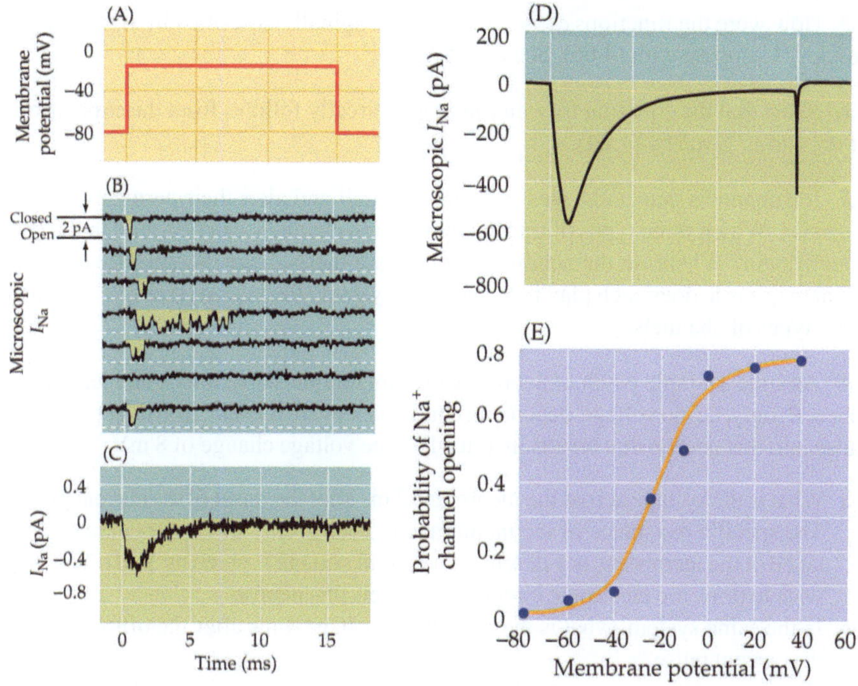

Fig. 1.15 Single ion currents and macroscopic current for a transient channel current. A. membrane voltage applied in a voltage-clamp experiment. B. individual ion currents. C. Mean ion channel current. D. Total current. E. Probability of channel opening (described by the steady state m_∞ activation function in the Hodgkin-Huxley equations) as a function of the membrane voltage. For the sodium channel, an additional probability as a function of the voltage describes the inactivation—not shown in this illustration, but compare this with Fig. 1.7, middle panel. Illustration from Neuroscience, 3rd edition (editors Purves et al.). Reproduced with permission from Oxford Publishing Limited

Problems

1.1 Estimate the Nernst potentials for sodium, chloride and calcium. Use values from Fig. 1.1 and (1.9).

1.2 Prove (1.14). Note that V_{rest} is given in (1.10).

1.3 Estimate how fast ion gradients will disappear if all ion pumps are stopped. Which currents are fast and slow in this respect? We will show in a later chapter that in reality ion gradients will not completely disappear, due to the Gibbs-Donnan effect.

1.4 Assume that $b = 0$, $m = 5/(5 + \exp(0.1 - 35 \cdot V))$ and $a = 4$ in (1.21). Show in a graph the resulting proportion of open channels as a function of the membrane voltage, V, in the range from -100 to $+100$ mV.

1.5 How were the functions $\alpha_n(V)$ and $\beta_n(V)$ originally described by Hodgkin and Huxley? Compare with (4.66). See e.g. [59].

1.6 Show that the equation for $\dot{n}$ in footnote 6 directly follows from the expressions of n_∞ and τ_n given by (1.24).

1.7 Ion channels detect changes in various stimuli and alter their permeabilities in response. Which three types of transmembrane ion channels can be found in the nervous system? Where on the neural membrane are these channels typically located? What major role does each play in the nervous system? Provide an example of two of these types of channels.

1.8 Assume a simple patch of neuron, with a membrane capacitance and leak resistance, R, only, i.e. no voltage-dependent conductances are present. You inject a current pulse into this neuron that results in a steady-state voltage change of 8 mV.

a What is the voltage across the membrane 2 ms after the onset of the current pulse? The specific resistance of the membrane is 2000 Ωcm^2 and the specific capacitance of the membrane is 1 μF/cm^2. The initial voltage is given by $V(0)$. You may wish to draw the electronic circuit that models this neuron.
b In the same system, what is the membrane voltage 4 ms after the offset (end) of the current pulse?

1.9 Find the total membrane current, I, and the membrane potential V_m at the peak of an action potential, given the following conductances (at this peak) and equilibrium potentials: $g_{Na} = 1.0$ mS/cm^2, $g_K = 0.15$ mS/cm^2, $E_{Na} = +55$ mV, $E_K = -58$ mV.

1.10 For a given neuron, the extracellular $[K^+]$ is 5 mM and the intracellular $[K^+]$ is 140 mM. Suppose the current carried by potassium ions is zero and the K^+ conductance is 0.5 mS, what is the membrane potential at 22 °C?

1.11 We stated in the text that the sodium current can be blocked using TTX. What is an alternative to limit the potential sodium current contribution to the total current in voltage clamp experiments, and was actually used by Hodgkin and Huxley, as well?

1.12 A deep sea exploration mission has just discovered a new species of squid and brought it to your lab. From the squids giant axon, you observe action potentials with a unique shape and have reason to believe that three ions are responsible for production of action potentials in this squid's giant axon: Na^+, K^+ and Ca^{2+}. Your goal is to measure Ca^{2+} conductances during these odd action potentials. Your lab has tetrodotoxin (TTX), KCl and instrumentation to perform voltage clamp experiments. Since this squid will expire soon, you have no time to grab other materials. How will you measure the Ca^{2+} contribution to these action potentials?

1.13 For decades after their description of action potentials, Hodgkin and Huxley presumed that neuronal membranes must contain voltage-sensitive ion channels that controlled selective ion permeability.

a What method provided the first *direct* evidence of the presence of transmembrane ion channels? What evidence did this method reveal?
b What are major differences between voltage-gated Na^+ and K^+ channels in response to depolarization?

1.14 Simulate a Hodgkin-Huxley neuron in Matlab; you could use the code `hh.m`. Reconstruct the curves shown in Fig. 1.7 and determine the ratios of the steady state conductances of the sodium and potassium currents as a function of the membrane voltage. Use $\overline{g}_K = 36$ mS/cm^2 and $\overline{g}_{\text{Na}} = 120$ mS/cm^2. If only these two channels were present, what would be the value of the resting membrane potential?

1.15 Inward currents (positive charge enters the cell) will depolarize the cell; outward currents will typically make the membrane potential more negative. In some situations, however, outward currents will (slightly) lower the membrane potential, but their functional role is inhibition. How is this possible?

1.16 In this exercise, you will calculate some biophysical characteristics associated with the charge separation across the thin cell membrane.

a Make an estimate of the field strength across the neuronal cell membrane. Assume typical values for its thickness and take realistic values of the membrane potential.
b Make an estimate of the surface charge on the cell membrane if the membrane potential is −70 mV.
c Make an estimate of the pressure resulting from the electrical forces exerted on the cell membrane.
d During generation of an action potential, the thickness of an axon changes approximately 10 Angstrom, and the temperature changes approximately 20 μK.[8] Can you make some suggestions why this will occur?

1.17 Explore the membrane potential described by the Hodgkin-Huxley equations for different ion gradients of potassium and sodium, using the code you used in Problem 1.14. Also, explore the effect of different values of the external current I. You could, for instance, plot the frequency of the action potentials as a function of the Nernst potentials of sodium and potassium.

a What do you observe if you would only increase the extracellular potassium concentration?
b In which clinical situation could such changes in Nernst potentials of sodium or potassium occur?

1.18 If you take your program from the previous exercise, you can also make some modifications in the channel gating, thus simulating a particular channelopathy. A clinical example is Dravet syndrome, a devastating childhood epilepsy disorder, resulting from abnormal voltage-gated sodium channels. Simulate this abnormal gating in your Hodgkin-Huxley neuron and explain your results in the context of Dravet.

---

[8] Tasaki et al. Biophysical Journal, 1989.

1.19 The potassium current as described in the HH-equations is a persistent current. This is not the case for all potassium currents. The A-type potassium current (I_A) shows relatively rapid inactivation and contributes to action-potential repolarization in cortical neurons. Modify the HH equations with this additional current and show that this current indeed limits the firing rate for a particular input.

Chapter 2
Synapses

> *Neurons like one another very much. They respond to one another's messages, so they basically chat all day, like people do in society.*
>
> — Rodolpho Llinás

Abstract We discuss how neurons transmit signals between neurons, focusing on the chemical synapse. After a more phenomenological treatise, we derive expressions to quantify synaptic transfer. A few neurological diseases, characterized by abnormal synaptic transmission, are discussed, too.

2.1 Introduction

In the previous chapter, we reviewed and discussed the generation of action potentials. To eventually realize function, however, neural transmission is essential. We will not treat the propagation of the action potential along dendrites and axons,[1] but focus on transmission of information between neurons. Neurons primarily communicate with each other via special structures: *synapses*, that actually come in two flavors: chemical synapses and electrical synapses, see Fig. 2.1.

Chemical transmission is controlled by Ca^{2+}-dependent release of neurotransmitters, while electrical transmission is mediated by intercellular channels: gap junctions. The gap junctions directly connect the cytoplasm of neurons, allowing transmission of various molecules and electrical impulses. Similar to the chemical

[1]In e.g. [[42, 65, 130]] action potential propagation is discussed.

Electronic supplementary material The online version of this chapter (https://doi.org/10.1007/978-3-662-61184-5_2) contains supplementary material, which is available to authorized users.

The original version of this chapter was revised: The belated corrections have been incorporated. The correction to this chapter is available at https://doi.org/10.1007/978-3-662-61184-5_11

[1]In e.g. [42, 65, 130] action potential propagation is discussed.

M. J. A. M. van Putten, *Dynamics of Neural Networks*,
https://doi.org/10.1007/978-3-662-61184-5_2

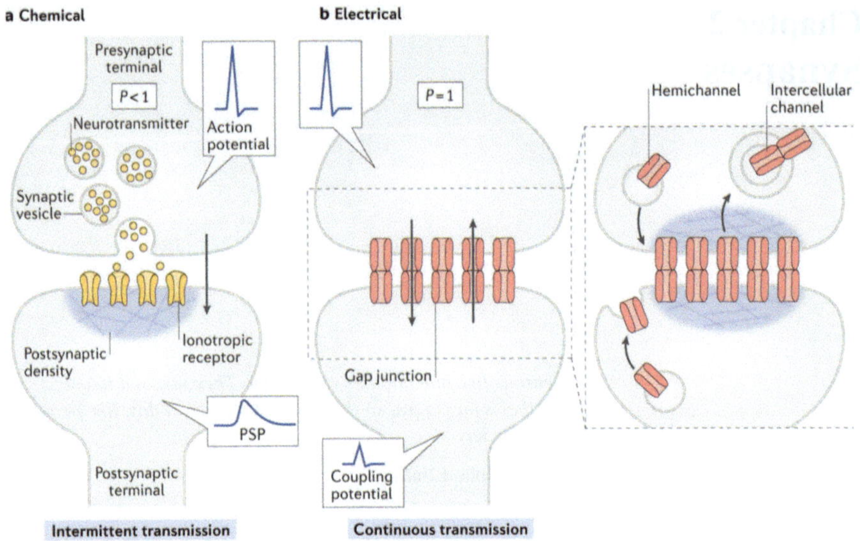

Fig. 2.1 The chemical synapse **a** is unidirectional, and transmission is mediated by the release of a neurotransmitter, that subsequently interacts with a ionotropic receptor/ ligand-gated ion channel at the postsynaptic site. The resulting change in membrane potential is the postsynaptic potential (PSP). The electrical transmission **b** is bidirectional and is mediated by the gap junctions that allow passage of electrical currents. The potential that is induced is known as the coupling potential. Gap junctions can be removed and inserted, thus controlling the strength of the transmission. At the chemical synapse, the release of a neurotransmitter is (in part) probabilistic ($P < 1$); at the gap junction, the transmission is completely defined ($P = 1$). Reprinted from [3], with permission from Springer Nature

synapse, the electrical synapse can also be modified and varied in strength [3]. Gap junctions are not restricted to neurons; many other cells contain gap junctions, too, including cardiomyocytes. If we further discuss synapses, we will implicitly assume that we are considering chemical synapses.

The chemical synapse was discovered by the Spanish anatomist, Santiago Ramón y Cajal (1852–1934), who used the staining techniques developed a few years earlier by the Italian anatomist, Camillo Golgi (1843–1926). These techniques allowed selective coloring of neurons, that would turn black, which allowed good visualization under the microscope. This black reaction, or the *reazione nera*, is still in use today. Golgi thought that neurons were being connected to each other, without any special structure, thus forming a large network, the *retinaculum*. Soon after, this proved wrong by the findings of Cajal.[2] The main findings Cajal published in 1890 were that a neuron is the structural and functional unit of the central nervous system and that neurons are *individual cells*, consisting of a cell body, axons and dendrites, where communication between neurons occurs via synapses. Both Golgi and Cajal were awarded the Noble prize for medicine or physiology in 1906.

[2]Remarkably, a lot of Cajal's research was actually performed in his kitchen, that mainly served as a laboratory.

Synapses are highly dynamic structures and can be viewed as the basic unit of neural circuit organization [96]. At the synapses, neurons excrete neurotransmitters, that serve as chemical signals for the receiving neuron. At first glance, a synapse may seem a simple connection that can either increase the likelihood of firing of the receiving neuron, an excitatory synapse, or decrease this likelihood, an inhibitory synapse. In fact, there is a lot of experimental evidence that this functional unit is very complex, and its properties may change over time, among others being a function of activity, either physiological or pathological. These changes in synaptic strength occur on various timescales, and include long- and short-term (synaptic) plasticity. Changes on a time scale of hours to days are known as long term depression (LTD) and potentiating (LTP), while short term plasticity has a much shorter time scale, typically of the order of 100–1000 ms. We will discuss some of these properties later in this chapter.

2.2 A Closer Look at Neurotransmitter Release

Let us have a closer look at the synaptic structures, where a cascade of events is necessary in order to release a neurotransmitter. In Fig. 2.2, we show an EM-photograph of a chemical synapse illustrating how synaptic vesicles fuse with the cell membrane to release a neurotransmitter.

The primary trigger for the release of a neurotransmitter is the arrival of an action potential, that travels from the soma towards the end of the axon, the axon terminal. Voltage sensitive (voltage-gated) calcium channels will now become activated by the changes in membrane potential caused by the arrival of this action potential, and a

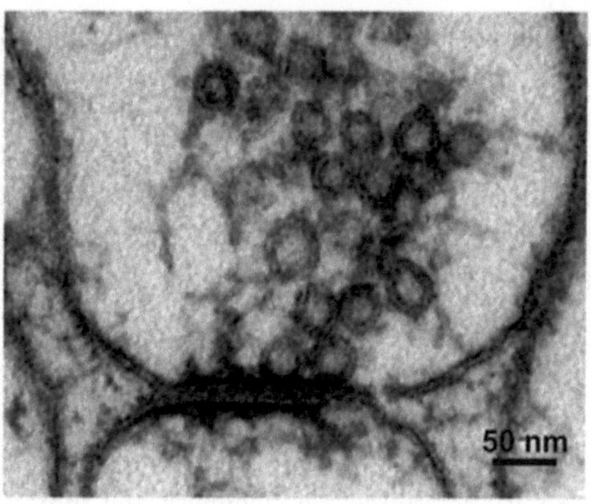

Fig. 2.2 Electron micrograph of a glutamatergic hippocampal synapse. Synapses can be readily identified by a heavily stained postsynaptic density in close proximity to presynaptic terminals filled with clusters of synaptic vesicles. Image from Guzman et al, Involvement of ClC-3 chloride/proton exchangers in controlling glutamatergic synaptic strength in cultured hippocampal neurons, Frontiers in Cellular Neuroscience, 8:142;2014

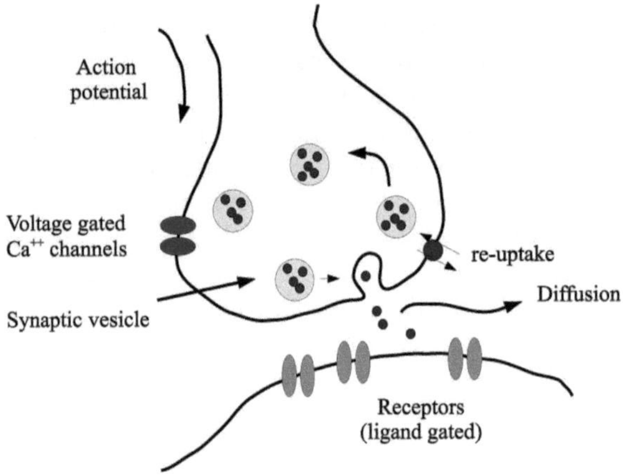

Fig. 2.3 Schematic of the processes that take place in a chemical synapse. After arrival of the action potential, synaptic vesicles fuse with the presynaptic membrane, releasing the neurotransmitter (•) that will interact with the postsynaptic ligand-gated ion channel. Note, that close to this channel, the voltage-gated channels are present. If the (graded) postsynaptic membrane potential change is sufficiently large, the voltage-gated channels will be activated, generating an action potential

short lasting increase in calcium conductance causes a large influx of calcium ions into the axon. This subsequently triggers the docking of synaptic vesicles, finally resulting in the release of the neurotransmitter into the synaptic cleft, summarized in Fig. 2.3. For a single action potential, an average of 0.5–1 synaptic vesicle is released [73]; average firing rates of cortical neurons range are about 0.1–1 Hz, but maximum rates can be up to 200 Hz. Peripheral motor neurons can generate action potential frequencies up to 40 Hz in situations where maximal muscle force is generated.

There exists a well-defined relationship between the change in presynaptic membrane voltage, V_{pre}, and the amount of neurotransmitter finally released. In physiological conditions, the presynaptic change in membrane voltage is the main determinant of the amount of neurotransmitter released. This allows us to formulate a simple expression that relates the in change presynaptic membrane voltage to the amount of neurotransmitter release. A good fit to the total amount of neurotransmitter (NT) release by a single action potential is given by the expression

$$[NT](V_{\text{pre}}) = \frac{NT_{\max}}{1 + e^{-(V_{\text{pre}} - V_p)/K_p}}, \tag{2.1}$$

where $NT_{\max}$ is the maximal concentration of neurotransmitter (NT) in the synaptic cleft and both K_p and V_p determine the steepness and threshold for the release. Good values for these constants are $V_p = 2$ mV and $K_p = 5$ mV. Typically, approximately 1 mM of transmitter is the maximum concentration released. Equation 2.1 thus provides a simple and smooth transformation between presynaptic voltage and transmitter concentration.[3]

2.3 Modeling Postsynaptic Currents

To model the events that occur at the postsynaptic junction, we have to take several variables into account: the number of neurotransmitter molecules released,[4] the time dependency of this release, the number of post-synaptic receptors, the affinity of the receptor for the transmitter, the rate of disappearance of transmitter from the cleft by diffusion, re-uptake (density of uptake transporters and kinetics of removal), or enzymatic breakdown and the conductance of a single channel. For instance, the AMPA receptor has a conductance of 8 pico Siemens (pS), while the NMDA receptor has a conductance of about 50 pS.

In the previous Sect. 2.2, we discussed how to model the *release* of neurotransmitter in the synaptic cleft, and we introduced (2.1). A further simplification to the transmitter release is that it occurs as a *brief pulse*: if an action potential arrives at the presynaptic axon, the neurotransmitter is released and reaches its maximum concentration in an infinitesimally short period. Therefore, after the arrival of an action potential at the presynaptic axon, we now assume that the release of the neurotransmitter is immediate, reaching its maximum concentration without any delay.[5] This greatly simplifies the (computational) modeling involved, as will become clear in the next subsection 2.3.1.

Similar to the description of the transmitter release, several approaches can describe the postsynaptic currents. Here, we will assume that we have a sufficient number of receptors available that we can model their *average behavior*. Remember, that channel opening will typically result in changes in transmembrane currents, I_{syn}, where the driving force is the difference between the membrane potential, V_m and the Nernst potential or reversal potential of the channel, E_{syn}, expressed as:

$$I_{\text{syn}} = g_{\text{syn}}(t)(V_m(t) - E_{\text{syn}}) \tag{2.2}$$

with $g_{\text{syn}}(t)$ the *time-dependent synaptic conductance*. In contrast to the voltage-gated channels, this conductance is not a function of the transmembrane voltage but

[3]More details can be found in: Methods in Neuronal Modeling, Edited by Koch, C. and Segev, I. (2nd Edition) MIT Press, Cambridge, 1998.

[4]A single vesicle contains between 500–10000 molecules, with different values for different neurotransmitters and the location within the central or peripheral nervous system. As an example: glutamatergic neurons: 3500 molecules Glu per vesicle; cholinergic neurons: 2000–9000 molecules ACh per vesicle.

[5]This refers to the delay between the action potential and the release of neurotransmitter. This is not the total delay between the action potential arriving at the presynaptic neuron and the generation of an action potential in the receiving neuron.

a function of the amount of neurotransmitter that is excreted and the density of the postsynaptic receptors (i.e. ligand gated ion channels).

Synapses are excitatory or inhibitory, determined by their reversal potentials. If the reversal potential, E_{syn}, is larger than a threshold membrane potential, $V_{\text{threshold}}$, activation of this synapse tends to excite the cell. The corresponding resulting potential is called an excitatory postsynaptic potential (EPSP). A synapse that produces a conductance increase and whose reversal potential is more negative than a threshold potential, tends to inhibit the cell from firing. The corresponding resulting potential is called an IPSP: an inhibitory postsynaptic potential. Therefore, synapses with $E_{\text{syn}} \leq V_{\text{rest}}$ are clearly inhibitory. Activation of these synapses will hyperpolarize the cell. It is possible that $E_{\text{syn}} \approx V_{\text{rest}}$. In that case, activation of the synapse will result in an increased conductance, without a change in membrane potential, a phenomenon known as *shunting inhibition* or silent inhibition. It is also possible, that a synapse increases the membrane voltage, V_m, but is still inhibitory. This holds for all synapses with $V_{\text{rest}} < E_{\text{syn}} < V_{\text{threshold}}$.

Our goal is to find an expression for the fraction or number of open channels as a function of time when neurotransmitter is released. Therefore, we wish to find an explicit expression for $g_{\text{syn}}(t)$ in (2.2).

2.3.1 The Synaptic Conductance

After a neurotransmitter or agonist, A, is released released from the presynaptic axon it will enter the synaptic cleft and diffuse towards postsynaptic receptors R where it can interact with the receptor R, creating the agonist-receptor complex, AR. We assume that the binding of the neurotransmitter with the ligand-gated receptor is reversible. The binding of the agonist with the receptor is described by the rate constant, k_1, (units $M^{-1}s^{-1}$) and the unbinding from the receptor by the unbinding rate, k_{-1} (units s^{-1}). The ratio of the binding and unbinding rate is given by the dissociation constant K_d

$$K_d = \frac{k_{-1}}{k_1} = \frac{[A][R]}{[AR]} \tag{2.3}$$

with [.] the equilibrium concentrations. The dissociation constant, K_d, indicates the strength of binding between A and R in terms of how easy it is to separate the complex AR. Therefore, if a *high* concentration of A and R is required to form AR, this indicates that the strength of binding is *low*. The K_d would therefore be *higher* as more of A and R are required to form AR. It follows that the smaller K_d, the stronger the binding.

After binding of the agonist with the ligand-gated channel, the channel is still *closed*, and it takes some time to subsequently open.[6] Opening of the channel after

[6] Recall that voltage-gated ion channels also open with a time delay in response to a voltage change, as discussed in the previous chapter.

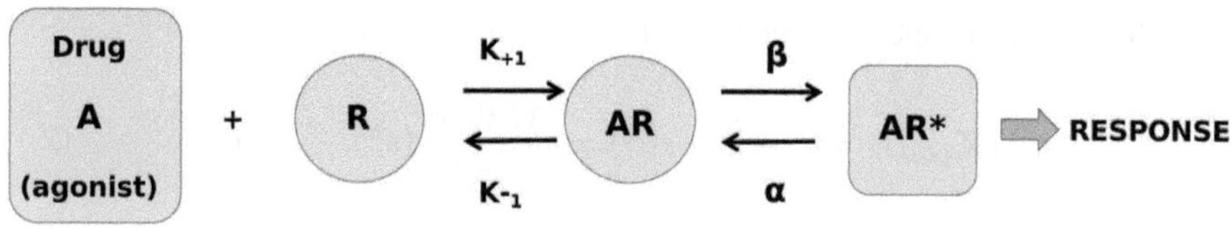

Fig. 2.4 State diagram of a ligand-receptor interaction. A is the neurotransmitter or agonist, R the free receptor, AR the transmitter-receptor complex (still closed) and AR* the open transmitter-receptor complex

the AR-complex is created is described by the mean channel opening rate, β (units s^{-1}). The larger the opening rate, β, the more channels open per time unit after formation of the AR complex. Channels do not remain open forever. Deactivation of the channels results from desensitisation.[7] This is analogous to the inactivation state of transient voltage-gated channels, and is described by the mean channel closing rate, α (units s^{-1}). This is summarized[8] in Fig. 2.4.

We can write these interactions as

$$A + R \underset{k_{-1}}{\overset{k_1}{\rightleftharpoons}} AR \underset{\alpha}{\overset{\beta}{\rightleftharpoons}} AR_{\text{open}}. \tag{2.4}$$

We will further assume that the binding of the neurotransmitter that is released to the receptor is instantaneous (a pulse) and that unbound neurotransmitter disappears (so no rebinding). Therefore, we can remove the k_1 in (2.4) to arrive at

$$A + R \underset{k_{-1}}{\leftharpoondown} AR \underset{\alpha}{\overset{\beta}{\rightleftharpoons}} AR_{\text{open}}. \tag{2.5}$$

Using this assumption allows us to start with a certain number of AR complexes. You may interpret this as if the vesicles from the presynaptic axon excrete the AR-complex immediately. Typical values for the constants are $k_{-1} = 30 - 500\ \text{s}^{-1}$, $\beta = 10^5 - 10^6\ \text{s}^{-1}$ and $\alpha = 350\text{–}1000\ \text{s}^{-1}$.

We can now write the differential equations that describe the change in AR and the change in the AR_{open}. This follows from (2.5), resulting in

$$\begin{aligned} \frac{dAR}{dt} &= -k_{-1}AR - \beta AR + \alpha AR_{\text{open}} \\ \frac{dAR_{\text{open}}}{dt} &= \beta AR - \alpha AR_{\text{open}}. \end{aligned} \tag{2.6}$$

---

[7]We use desensitisation here in the general sense. Some authors use desensitisation for long term channel inactivation effects (minutes or more) and reserve the word inactivation for short term (seconds or less).

[8]This state diagram is not generic for all ligand-receptor interactions, but it is sufficient for our discussion. For instance, it is also possible that the activated agonist-receptor complex, AR*, dissociates in A and R, or that a receptor has multiple binding sites for a ligand, as is for instance the case of the acetylcholine receptor that has two binding sites for acetylcholine.

The solution for the open channels as a function of time is by[9]

$$AR_{\text{open}}(t) = c\left(e^{\lambda_1 t} - e^{\lambda_2 t}\right) = \frac{AR(0)\beta}{\lambda_1 - \lambda_2}\left(e^{\lambda_1 t} - e^{\lambda_2 t}\right) \tag{2.7}$$

where we used the initial condition that at t=0 there are no open channels, $\text{AR}_{\text{open}}(0) = 0$. We further assumed that there is instant binding of agonist with the receptor at $t = 0$, resulting in an initial number of agonist-receptor complexes AR(0). For the synaptic conductance, it now holds that it equals the product of the single channel conductance times the number of open channels:

$$g_{\text{syn}}(t) = AR_{\text{open}}(t) \cdot \text{single channel conductance.} \tag{2.8}$$

For example, when the number of open channels at a particular moment is 80 and the single channel conductance is 10 pico Siemens (pS), then $g_{\text{syn}} = 80 \cdot 10 = 800$ pS. Further details can be found in e.g. [15, 65].

Conductances can also be related to the surface area of a cell, and are then expressed as conductance densities, with unit S/m^2. Setting $\lambda_1 = -1/\tau_1$ and $\lambda_2 = -1/\tau_2$, we can alternatively write

$$AR_{\text{open}}(t) = \frac{AR(0)\beta\tau_1\tau_2}{\tau_1 - \tau_2}\left(e^{-t/\tau_1} - e^{-t/\tau_2}\right) \tag{2.9}$$

with $\tau_1 > \tau_2$. We replace the product of $AR(0)\beta$ and the single channel conductance by Ag_{max} to arrive at

$$g_{\text{syn}}(t) = \frac{Ag_{\text{max}}}{\tau_1 - \tau_2}\left(e^{-t/\tau_1} - e^{-t/\tau_2}\right), \text{ for } \tau_1 > \tau_2, \tag{2.10}$$

where A is a normalization constant chosen so that g_{syn} reaches a maximum value of g_{max}. Typical values for g_{max}, for a single synaptic input, are 0.1 to 1 nS.

Another normalization often used in the literature is to normalize the integral of the synaptic response, using $\int_0^\infty g_{syn}(\tau)\,d\tau = \bar{g}_s$. We added the subscript s to differentiate from the $\bar{g}$ we introduced in the previous chapter as the mean voltage-gated channel conduction with unit S/cm^2. This results in

$$g_{\text{syn}}(t) = \frac{\bar{g}_s}{\tau_1 - \tau_2}\left(e^{-t/\tau_1} - e^{-t/\tau_2}\right). \tag{2.11}$$

Check that indeed $\frac{1}{\tau_1 - \tau_2}\int_0^\infty (\exp(-t/\tau_1) - \exp(-t/\tau_2) d\tau = 1$. The constant $\bar{g}_s$, has units S$\cdot s$ (Siemens x seconds).

The time constants, $\tau_1 > \tau_2$, control the rising and falling phases of synaptic conductance. The maximum value of the conductance is reached if

[9]The two equations (2.6) are a planar system of autonomous linear differential equations. In Chap. 4, Example 4.2, we will show a method to solve it.

$$t = t_{\max} = \ln(\tau_1/\tau_2)\frac{\tau_1\tau_2}{\tau_1 - \tau_2}. \tag{2.12}$$

Both (2.10) and (2.11) for the synaptic conductance are common in the neuroscience community. Depending on the characteristics of the particular ligand gated channel, the two exponentials functions can be replaced by a single exponent with one time constant. This is presented next.

2.3.2 Very Fast Rising Phase: $\tau_1 \gg \tau_2$

In the special case that $\tau_1 \gg \tau_2$, (2.10) reduces to

$$g_{\text{syn}}(t) = g_{\max} e^{-t/\tau} \tag{2.13}$$

and (2.11) to

$$g_{\text{syn}}(t) = \frac{\bar{g}_s}{\tau} e^{-t/\tau} \tag{2.14}$$

with $\tau \approx \tau_1$. This expression holds for synapses where the rising phase is orders of magnitude larger than the falling phase, and the maximum conductance, $g_{\max} = \frac{\bar{g}_s}{\tau}$ is reached at $t = 0$; you can also check that this holds by taking the limit $\tau_2 \to 0$ of (2.12).

2.3.3 Equal Time Constants: $\tau_1 = \tau_2$

There is another special situation, where the time constants τ_1 and τ_2 are the same. It can be proven[10] that (2.10) then reduces to

$$g_{\text{syn}}(t) = \frac{g_{\max} t}{\tau} e^{(1-t/\tau)} \tag{2.15}$$

and (2.11) to

$$g_{\text{syn}}(t) = \frac{\bar{g}_s t}{\tau^2} e^{(-t/\tau)}. \tag{2.16}$$

This function increases rapidly to a maximum value $g_{\max} = \bar{g}/\tau e$ at $t = \tau$, and following its peak the conductance decreases more slowly to zero. A synapse that can be modeled in this way is also known as an *alpha synapse* and the (2.15)–(2.16) are known as *alpha functions*. In Fig. 2.5 the three functions that describe the synaptic conductance are illustrated.

[10]The proof is left to the reader.

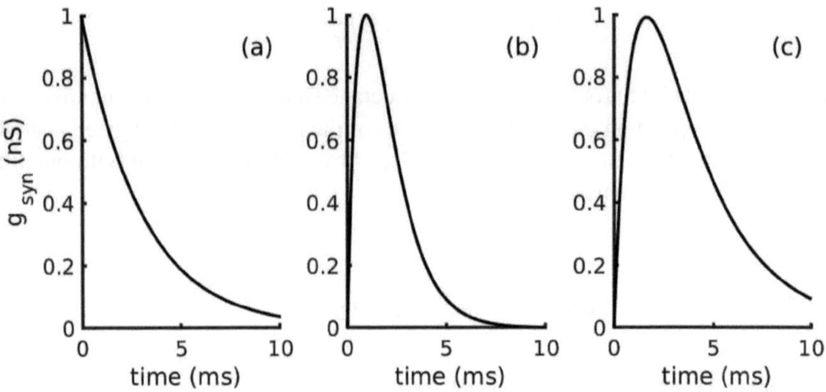

Fig. 2.5 Time course of synaptic conductance g_{syn} for the three functions discussed. **a** Very fast rising phase with $\tau = 3$ ms. **b** alpha function with $\tau = 1$ ms. **c** dual exponential with $\tau_1 = 3$ ms and $\tau_2 = 1$ ms. In all three cases, the maximum conductance, $g_{max} = 1$ nS

2.4 Channelopathies

Proper gating of ions is critical for normal function and several diseases exist that find their origin in an abnormal function of ion channels: channelopathies. Diseases include epilepsy, movement disorders, migraine, myotonia and congenital insensitivity to pain (Table 2.1). Sodium channelopathies are quite common and in many of

Table 2.1 Examples of neurological channelopathies. The total number of channelopathies associated with neurological disorders is much larger. Channelopathies include abnormal gating of voltage-gated and ligand-gated channels, both in the central and peripheral nervous system

Channel	Gene (channel)	Disorder
Na^+	SCN1A (α-subunit of $Na_v1.1$)	Generalised epilepsy with febrile seizures (GEFS+)
		Intractable epilepsy with generalised tonic-clonic seizures (IEGTC)
	SCN2A (α-2 subunit of $Na_v1.2$)	Benign familial neonatal infantile seizures (BFNIS), Infantile spasms, GEFS+
	SCN4A ($Na_v1.4$)	Dystrophic myotonia
	SCN9A ($Na_v1.9$)	Congenital insensitivity to pain
K^+	KCNQ1	Benign familial neonatal convulsions
	KCNA2 ($K_v1.2$)	Myoclonic epilepsy and ataxia
Ca^{2+}	CACNA1A ($Ca_v2.1$ channel α-subunit	Episodic ataxia and childhood absence epilepsy
		Familial hemiplegic migraine (FHM1)
GABA	GABRA1 (α-subunit of GABA receptor)	Childhood absence epilepsy
AChR	α-subunit of AChR (CHRNA1)	Myasthenic syndromes

these disorders the opening time is prolonged which is caused by an abnormal inactivation. An example is hyperkalaemic periodic paralysis (hyperPP) that is caused by abnormal voltage-gated sodium channels in the myocytes (Fig. 2.6). Patients experience muscle hyperexcitability or weakness that is exacerbated by potassium, heat or cold. The clinical name may be confusing, as it is called hyperkalaemic periodic paralysis. The name results from the fact that an increased serum potassium is often the trigger for the clinical scenario. Recall that an increase in serum potassium will tend to reduce the membrane potential (make it less negative), which will activate the (abnormal functioning) voltage-gated sodium channels, triggering the initial hyperexcitability. This may subsequently evolve to weakness from a persistent depolarisation of the motor end-plate.

2.5 Synaptic Plasticity

The expressions just derived suggest that synapses are in essence static: for a given amount of neurotransmitter that is excreted in response to a presynaptic pulse train and interacts with the postsynaptic receptor, the current induced is the same. This, however, is not the case: the postsynaptic current may change, reflecting synaptic plasticity. This is a very important characteristic: you would not remember anything from studying this text if your synapses did not change! Major contributions to the structural mechanisms involved in memory storage were made by Erik Kandel [5, 64]. In 2000 he received the Nobel Prize in Physiology or Medicine for his discoveries of how the efficiency of synapses can be modified, and which molecular mechanisms are involved. As an experimental model, he used the nervous system of a sea slug (Aplysia). In studying the gill-withdrawal reflex of the Aplysia, he discovered that protein phosphorylation in synapses is essential for learning and memory. In various forms of pathology, synaptic transmission is changed, as well. Examples include metabolic stress, such as hypoxia [55] and Alzheimer's disease [61]. We will discuss the effects of hypoxia on synaptic function in more detail in Chap. 8.

The mechanisms involved in changing the synaptic efficacy (strengthening or weakening) are varied, and include changes in the probability of neurotransmitter release, insertion or removal of post-synaptic receptors and phosphorylation and dephosphorylation. Time scales involved in synaptic plasticity are varied, and often dichotomized in short and long term plasticity. Short-term synaptic plasticity (STP) acts on a timescale of tens of milliseconds to a few minutes, and long-term plasticity (LTP) on time scales from minutes to months and even longer.

2.5.1 Short Term Synaptic Plasticity

Two types of short term plasticity have been observed in experiments: Short-Term Depression (STD) and Short-Term Facilitation (STF). STD results from depletion

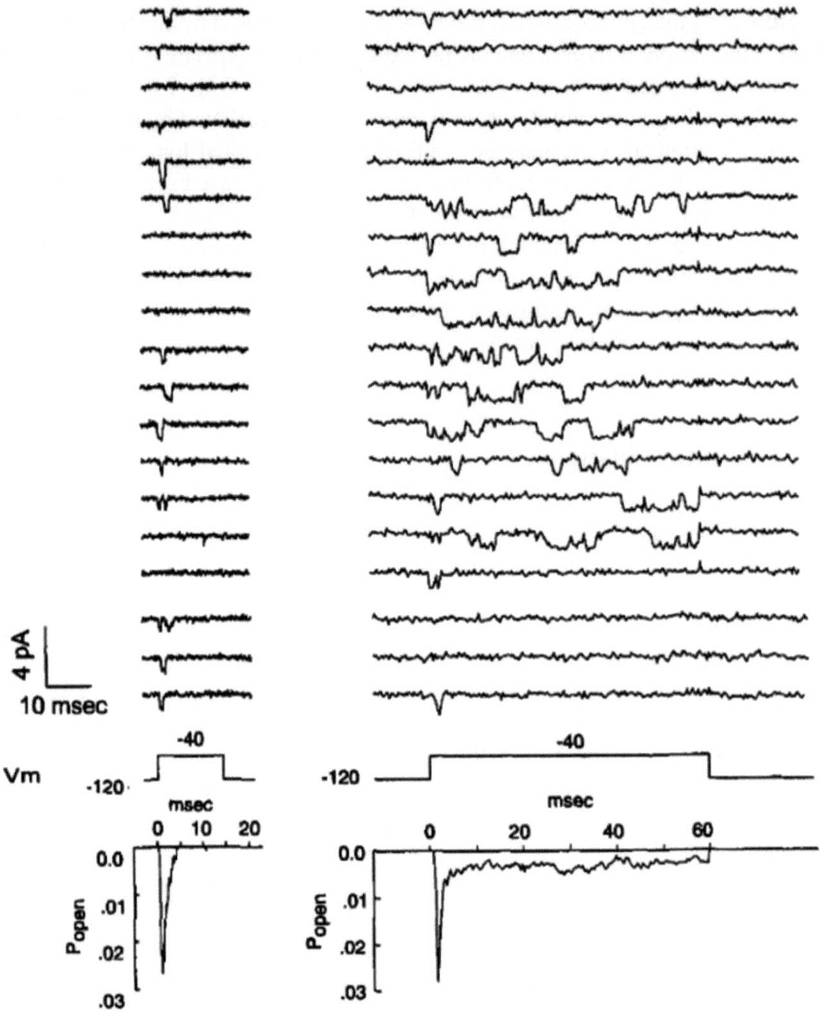

Fig. 2.6 Impairment of inactivation of muscle sodium channels in hyperkalaemic periodic paralysis (HyperPP). Na-currents were elicited with depolarizing pulses in cell-attached patches on normal (left) and HyperPP (right) myotubes. The latency to opening and the current amplitude were unchanged, but inactivation was abnormal, as evidenced by reopenings (downward current deflection) and prolonged open times. The clustering of noninactivating behavior in consecutive trials suggests a modal switch in gating. Ensemble averages (bottom) show the increased steady-state open probability caused by disruption of inactivation. Adapted from Cannon et a1, Sodium Channel Defects in Myotonia and Periodic Paralysis. Annu. Rev. Neurosci. 1996:19; 146–164. Reprinted with permission of Annual Reviews, Inc.

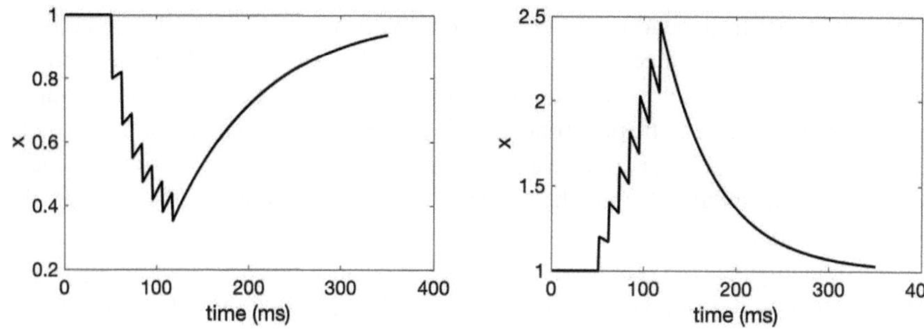

Fig. 2.7 Left: simulation of short-term depression, with $p = 0.6$ and $\tau = 100$. Right: short-term facilitation. Values taken were $p = 1.2$ and $\tau_x = 60$. Note that the change in x is larger when its value differs more from 1, according to (2.18)

of neurotransmitters during synaptic signaling; STF is caused by influx of calcium into the axon terminal after spike generation, resulting in an increased release of neurotransmitters for the next spike.

A straightforward way to phenomenologically model STP is suggested by Izhikevich [59]: if a presynaptic neuron fires, the maximal synaptic conductance is multiplied with a factor x. This scalar factor is subsequently multiplied with $p > 0$, reflecting STF, and $p < 0$ STD, at all times when a presynaptic neuron fires:

$$x \Rightarrow p \cdot x \text{ when a presynaptic neuron fires.} \tag{2.17}$$

As the changes in synaptic conductance are temporary (short term), synaptic efficacy returns to the baseline value, say $x = 1$, with a time constant τ_x. Therefore, we write for the change in x at all times that no presynaptic neuron fires

$$\dot{x} = (1 - x)/\tau_x. \tag{2.18}$$

This is illustrated, both for short-term depression and facilitation, in Fig. 2.7. Short term depression and facilitation result in a frequency dependent synaptic transmission: STD-dominated synapses favor information transfer for low firing rates, since high-frequency spikes rapidly deactivate the synapse (cf. Fig. 2.7 left). Contrary, STF-dominated synapses tend to optimize information transfer for high-frequency bursts, which increase the synaptic strength, as illustrated in Fig. 2.7 right.

2.5.2 *Long-Term Synaptic Plasticity*

Learning and memory storage depend on longer lasting changes in synaptic connections [1], that may result from correlation-based plasticity: synaptic inputs may change in strength, depending on correlations with pre- and postsynaptic activity. This concept was introduced by Donald Hebb, and is known as Hebbian learning. Hebb also stated that the presynaptic neuron should fire just before the postsynaptic neuron for the synapse to be potentiated. This process is known as spike time dependent plasticity (STDP): synaptic strength is increased if a synaptic events precedes the postsynaptic spike, whereas the synaptic strength is depressed if it follows the postsynaptic spike [83].

Persistent changes in synaptic strength are known as long term potentiation (LTP) and long term inhibition, where spike time dependent plasticity is one of the mechanisms involved. Other mechanisms to induced persistent changes in synaptic strength exist, as well. In non-Hebbian LTP, for instance, the order of pre- and postsynaptic events does not appear to be critical [127].

The phenomenon of LTP was first reported in 1973 by Bliss, Lomo and Gardner-Medwin. In their experiments in rabbits with high frequency (tetanic) stimulation (HFS) an enduring increase in the size of the synaptic potentials was found. This effect lasted for hours to months. This long-term synaptic plasticity is essential for learning and memory. About 20 years later, it was discovered that the precise timing of pre- and postsynaptic activity (spike-timing-dependent plasticity) is essential to induce LTP or LTD. At the molecular level, spike-timing-dependent plasticity is mediated by NMDA receptors that allow the timing of presynaptically released glutamate and the postsynaptic depolarization by allowing calcium influx into a dendritic spine [97]. A phenomenological description of the change in synaptic strength can be described as

$$\frac{\Delta g}{g} = a \exp^{-\beta\tau}, \tag{2.19}$$

with $\tau = t_{\text{post}} - t_{\text{pre}} > 0$ and a a constant. When $\tau < 0$ the synaptic strength will decrease, expressed as

$$\frac{\Delta g}{g} = -a \exp^{\beta\tau}. \tag{2.20}$$

2.6 Summary

In this chapter, we discussed chemical synapses and the interaction of neurotransmitters with ligand-gated receptors. We have derived expressions that describe the time-dependent behavior of the synaptic conductance, $g_{\text{syn}}(t)$. This allows us to study in a more quantitative matter normal and abnormal function that may occur at the level of the synapse.

Problems

2.1 Is it possible that an inhibitory synapse results in an increase in the membrane potential (the potential becomes less negative)? Explain your answer.

2.2 Derive (2.12), expressing the time at which g_{syn} reaches its maximum value.

2.3 We discussed that three different expressions can be formulated for synaptic transmission. Prove that for the alpha synapse, given in (2.16), it holds that the maximum conductivity is reached at $t = \tau$.

2.4 In Fig. 2.8 an equivalent circuit of a part of the nerve membrane is shown.

a. Modify the circuit shown in Fig. 2.8 to incorporate a time-dependent conduction change induced by activation of a synapse, i.e. sketch it.

To study the effect of voltage changes due to changes in synaptic conductance, we will make a further simplification by ignoring the voltage-dependent characteristics of the sodium and potassium channels, and setting their Nernst potentials to zero (you simply remove the voltage sources and the voltage dependent conductances). Also, set the Nernst potential for the leak current to zero. You should keep your synaptic battery and conductance, however.

b. Show that we can now write for the sum of the capacitative current, the current through the "leak channels" and the synaptic current, I_{syn}:

$$I_C + I_{\text{rest}} + I_{\text{syn}} = 0.$$

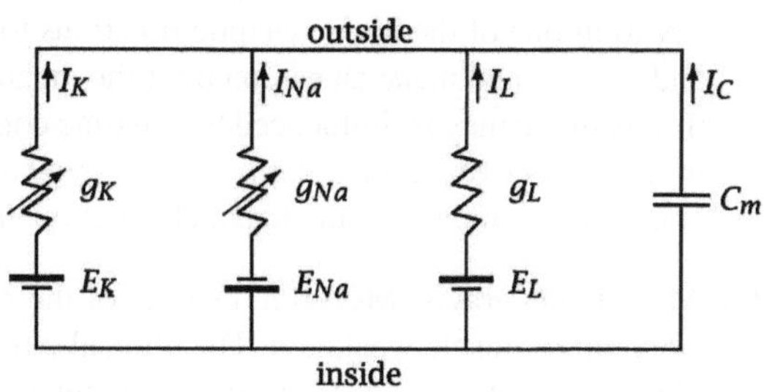

Fig. 2.8 Electrical equivalent circuit of a part of the cell membrane. The membrane capacitance is represented by a capacitor, C_m, the leak current conductance is g_L with corresponding Nernst potential E_L. Nernst potentials and voltage-dependent conductances for sodium and potassium are represented by g_i and E_i with $i =$Na, K for sodium and potassium, respectively

Alternatively,

$$C_m \frac{dV_m}{dt} + g_{\text{leak}} V_m + g_{\text{syn}}(t)(V_m - E_{\text{syn}}) = 0.$$

c. Now find an expression for the time constant of the cell membrane, assuming a rectangular $g_{\text{syn}}(t)$, during the opening and closing of the postsynaptic channels. Set $V_m(0) = 0$. Note, that this is the most simple "model" for a synaptic current, and contrast this with our previous discussion, that ultimately resulted in the single exponential or the sum of two exponentials to model the synaptic transmission.
d. Contrast the situation where a synapse opens and closes with the situation where you apply an external current to the cell model. Plot both situations in one graph (e.g. using Matlab, but you may sketch it by hand, too).

2.5 Each of the three functions for the synaptic conductance, g_{syn}, (2.10)–(2.16) can be used to model particular ligand-gated receptors. Provide some examples from the literature for particular neurotransmitters and their receptors.

2.6 Fast synapses may result in slow post-synaptic changes in the membrane potential. Explain how this is possible.

2.7 Use the experimental data recorded in a healthy volunteer at baseline and after injection with botulinum neurotoxin, at day 36. Recordings show miniature endplate potentials (MEPP) recorded in the extensor digitorum brevis muscle. Data are sampled with 4 kHz, with analog filter settings 20 Hz–5 kHz. For more details about these measurements, see e.g. [90]. Use the data files: MEPP0.asc and MEPP36.asc.

a. Make a plot showing the MEPPs at day 0 (baseline) and at day 36 after BTx injection. Describe your findings.
b. Try to fit one of the three synaptic functions to these experimental data at day 0 and day 36, and make an estimate for the mean channel open time. Realize, that this estimate may be influenced by the time constant of the cell membrane of the myocyte, as well. In fact, if $\tau_m > \tau_{\text{channel}}$ you are estimating the membrane time constant τ_m rather than the mean channel open time.

2.8 Myasthenia gravis (MG) is a disorder of the neuromuscular junction, resulting in (fluctuating) muscle weakness. The clinical manifestations may be limited to the face (bulbar myasthenia) or be generalized: generalized myasthenia gravis.

a. Explain in more details the nature of the synaptic dysfunction in MG. How would you model the time-dependent synaptic function in this disease?

In diagnosing MG, an ice pack test can be used in those patients presenting with a ptosis [27]. In this test, a small ice pack is placed above the symptomatic eye for about 2 minutes, and the change in ptosis is observed.

b. Explain why this is a sensitive test for MG.
c. Explain why inhibitors of acetylcholinesterase are beneficial in treating myasthenia. Relate this to the time-dependent synaptic function. What changes in the synaptic transfer function in this situation?

Part II
Dynamics

Chapter 3
Dynamics in One-Dimension

Change is most sluggish at the extremes precisely because the derivative is zero there.

— Steven Strogatz

Abstract In this chapter we discuss differential equations of one dependent variable. We treat fixed points and define rules to define the stability of these fixed points for linear and nonlinear equations, complemented by geometric reasoning. In the second part of this chapter, we discuss bifurcations: the change of the dynamics in response to a change in a control parameter.

3.1 Introduction

In the previous chapters we discussed the generation of the resting membrane potential, action potentials, including the Hodgkin-Huxley equations, and synaptic transmission. In this and the next chapter, we will build further on this understanding and show how various sorts of *dynamic behavior*, a key characteristic of biological systems, can arise from single and interacting neurons.

Dynamic behavior can loosely be defined as a change in a particular characteristic as a function of time, for instance a fluctuating membrane potential or the generation of oscillations. In biological systems, this is present across many spatial and temporal scales. Examples include the generation of an action potential (a change in membrane voltage at spatial scales of 1–10 μm and temporal scales of a few ms) in response to an excitatory input, the opening and closing of ligand gated channels by interacting with a neurotransmitter, rhythmic behavior of neuronal assemblies (spatial scales of several mm and temporal scales of 1 ms to hours), heart rhythms (temporal scales of seconds), breathing (temporal scale of minutes) and diurnal (24 h) sleep-wake behavior of a whole organism. In several diseases a change in dynamics is involved.

M. J. A. M. van Putten, *Dynamics of Neural Networks*,
https://doi.org/10.1007/978-3-662-61184-5_3

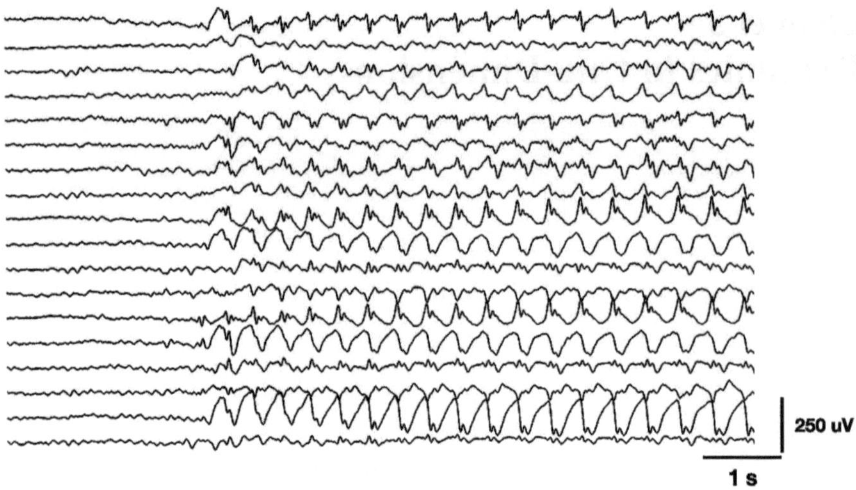

Fig. 3.1 Example of an EEG recording from a 16-y old patient with absence epilepsy, showing the sudden occurrence of generalized epileptiform discharges (3 Hz spike-wave complexes). A relevant clinical question is how this transition can occur. For instance, the system may possess intrinsic *bistability* or the change could result from a *bifurcation*. These concepts will be discussed in more detail in this and the next chapter

Examples include migraine, manic-depressive disorders and epilepsy. In these conditions, the brain 'switches' between the physiological and pathological state. This may occur either more gradually (hours-days), as in migraine or manic-depression, or very fast (seconds to minutes), as in seizures (Fig. 3.1). In patients with a manic depressive disorder, the transitions may also quasi-periodically alternate between two pathological conditions, as in manic-depressive disorders. Clinically relevant questions include why these transitions occur, which control parameter[1] is responsible for the change, and if the transitions to the pathological state can be prevented or predicted. Seizure prediction has recently regained interest, see for instance [68]. Dynamics is not limited to biological systems. In fact, it was originally a branch of physics, starting when Newton invented differential equations to describe his laws of the dynamics of motion and gravitation. Newton also solved the two-body problem, in particular the movement of the earth around the sun. Dynamics is also involved if two species interact, one as a predator and the other as prey. The interactions between the two populations can be described by the predator-prey equations (also known as the Lotka-Volterra equations). Here, we may observe dynamics in the population sizes of the predators and the prey, to be discussed in more detail in Chap. 4.

To better understand and describe the evolution of systems in time, be it biological or non-biological, we need mathematical tools, in particular differential equations.

[1] It can be useful to make a distinction between parameters and variables, where "a parameter is a variable that changes so slowly compared with other variables in the system that we can consider it as constant" [76].

Table 3.1 Nomenclature for differential equations from the number of dependent and independent variables. The example given for the PDE is the heat equation

	1 dependent variable	> 1 dependent variable
1 independent variable	Ordinary differential equation (ODE) e.g. $\frac{dy}{dx} = \frac{-x}{y}$	Systems of ODEs
> 1 independent variable	Partial differential equation (PDE) e.g. $\partial u/\partial t = \partial^2 u/\partial x^2$	Systems of PDEs

Although we did encounter these already in the previous chapters, in this and the next chapter we will discuss these equations in more detail, and illustrate why such equations are a necessary tool to more profoundly understand some of the rich dynamics of biological systems.

We emphasize that the more complex (and often interesting) a system is, the more complicated the differential equations needed to describe it. Globally, we can define two major characteristics: (i) the number of variables (the *dimension* of the system) and (ii) the presence of nonlinearity in the equations. In this chapter, we will focus on *single-variable* first order (n = 1) differential equations. Recall, that a differential equation is essentially a recipe describing the *change* of a variable of interest. In single variable differential equations, the recipe is only concerned with one dependent variable, a scalar; this is why such equations are also called *scalar* differential equations. It is also possible that the dynamics is characterized by more than a single dependent variable that may change. In those situations we need a *system* of differential equations. We further limit ourselves to *ordinary* differential equations (ODE), where the dependent variable is a function of a single independent variable. This is contrasted with partial differential equations (PDE), where the dependent variables are a function of more than one independent variables, for instance both time and space, as in the heat equation. An overview of the nomenclature is presented in Table 3.1. In this chapter we will be concerned with ODEs. In Chap. 4, we will discuss systems of ODEs. A further differentiation concerns linear and nonlinear equations and autonomous and non-autonomous equations. We will present examples of these categories. First, we will recapitulate some basics of ODEs, including analytical methods[2] to solve ODEs. If you are already familiar with these aspects, you can proceed to Sect. 3.3.

[2]An analytical solution can be defined as an expression that exactly describes the relation for the variable of interest to the parameters involved and (in principle) allows an exact solution. Some exclude limits and integrals from an analytic solution. If an analytic expression only involves addition, subtraction, multiplication, division, exponentiation to a rational exponent and rational constants it is generally referred to as an algebraic solution.

3.2 Differential Equations

Differential equations describe the evolution of systems as a function of the dependent variable, for instance time. In this section, we will very shortly review some general aspects of differential equations. For a more in depth treatment, students are referred to standard texts.[3]

3.2.1 *Linear and Nonlinear Ordinary Differential Equations*

A linear ordinary differential equation is any differential equation that can be expressed as

$$a_n(t)x^n(t) + a_{n-1}x^{n-1}(t) + \cdots + a_(t)x'(t) + a_0(t)x(t) = g(t). \tag{3.1}$$

The coefficients $a_0(t), \ldots, a_n(t)$ can be (non-)zero, (non-)constant or (non-)linear functions. What matters for the definition of linearity is the function $x(t)$ and its derivatives. If a differential equation cannot be written in the form of (3.1) it is a nonlinear differential equation. Nonlinear terms include products, powers, and functions of x, for instance $\sin(x)$. The order of the differential equation is defined by the largest derivative. For example $ax'' + bx' + cx = g(t)$ is a second order linear differential equation and $\sin(x)x''' = (1 + x)x' + x^2$ is a nonlinear, third order differential equation. We will limit our treatise to first order differential equations.

3.2.2 *Ordinary First-Order Differential Equations*

Ordinary first-order differential equations have the form

$$\frac{\mathrm{d}x}{\mathrm{d}t} = f(x, t). \tag{3.2}$$

It essentially tells you that the rate of change $\frac{\mathrm{d}x}{\mathrm{d}t} = \dot{x}$ is given by a particular rule defined by the function $f(x, t)$. In this case, the rule says that if the current value is x the rate of change is a function of both x and t, where the independent variable t represents time, and the dependent variable x corresponds to some dynamical physical quantity, for instance the membrane potential. An example is

$$\frac{\mathrm{d}x}{\mathrm{d}t} = \frac{x}{(t+1)^2} \tag{3.3}$$

[3] A very good introduction is "Elementary differential equations" [16].

with initial condition $x = x_0$ at $t = 0$. Recall that this is a linear, first order, differential equation as x appears to the first power, only. The solution of this equation is $Ce^{-1/(t+1)}$. You can solve this equation by yourself, but you can also use the solver in Matlab.

3.2.2.1 First-Order Autonomous Differential Equations

If there is no explicit dependence on the independent variable, the system is called autonomous. If the independent variable is time, such systems are also known as *time-invariant*. As we will always consider time as the independent variable, autonomous and time-invariance can be used interchangeably. The general expression for a first order autonomous differential equation is

$$\frac{\mathrm{d}x}{\mathrm{d}t} = f(x). \tag{3.4}$$

For autonomous differential equations, therefore, the rule does not explicitly depend on the time. It only cares about the current value of the variable x (which on its turn will depend on time). An example is radioactive decay

$$\dot{x} = -ax \tag{3.5}$$

with $a > 0$. The solution is given in Example 3.3.

3.2.3 *Solving First-Order Differential Equations*

Given a differential equation, we often wish to solve the equation, with the primary goal to get information about how the dynamics of the system behaves. However, most differential equations cannot be solved analytically and even numerical methods are not always sufficient. However, for a subclass of differential equations, analytical strategies do exist, to be discussed shortly. Another approach that we will discuss in more detail is to extract information about the solutions *without actually solving the differential equations* by geometric reasoning, discussed in Sect. 3.3.

3.2.3.1 Solving First-Order Non-autonomous Separable Differential Equations

The separation of variables method can be straightforwardly extended to a special class of scalar non-autonomous differential equations. A *separable* first order ordinary differential equation has the form

$$\frac{\mathrm{d}x}{\mathrm{d}t} = f(x)g(t), \tag{3.6}$$

where both f and g are smooth real-valued functions. In this case, separating and integrating yields the implicit equation

$$\int_{x(0)}^{x(t)} \frac{1}{f(u)}\,\mathrm{d}u = \int_{0}^{t} g(s)\,\mathrm{d}s, \tag{3.7}$$

where we used dummy variables of integration to avoid confusion.

Example 3.1 Consider the equation

$$\dot{x} = \frac{dx}{dt} = x^2 \sin(t). \tag{3.8}$$

As t appears explicitly, this is a non-autonomous equation, with $f(x) = x^2$ and $g(t) = \sin(t)$. We can write this as

$$\begin{aligned} \int \frac{dx}{x^2} &= \int \sin(t)dt = \\ -\frac{1}{x} &= \cos(t) + k \end{aligned} \tag{3.9}$$

or

$$x(t) = \frac{-1}{k + \cos(t)} \tag{3.10}$$

with k a constant.

* * *

3.2.3.2 Solving First-Order Non-autonomous Linear Differential Equations

Another class of non-autonomous equations for which we can derive a general solution formula are equations that are *linear* in the dependent variable x. The general form of this class of equations is given by

$$\frac{\mathrm{d}x}{\mathrm{d}t} = g(t)x + h(t), \tag{3.11}$$

where both g and h are continuous functions. Linear differential equations can be solved by a method called *variation of constants*. The idea is to first solve the special case $h(t) = 0$, the so-called *homogeneous* equation

$$\frac{dx}{dt} - g(t)x = 0. \tag{3.12}$$

Equation 3.12 is separable, so we can easily derive its general solution given by

$$x(t) = x(0)e^{G(t)}, \qquad \text{where} \qquad G(t) = \int_0^t g(s)\,ds. \tag{3.13}$$

Since we have not specified an initial condition, $x(0)$ in (3.13) is an arbitrary constant. To satisfy the *inhomogeneous* equation (3.11), we try an ansatz where we replace this constant by a time-varying function c, i.e

$$x(t) = c(t)e^{G(t)}. \tag{3.14}$$

Substituting (3.14) into (3.11) now results in a differential equation for our unknown function c, given by

$$\frac{dc}{dt} = e^{-G(t)}h(t), \qquad c(0) = x(0). \tag{3.15}$$

Direct integration of (3.15) then yields

$$c(t) = x(0) + \int_0^t e^{-G(s)}h(s)\,ds. \tag{3.16}$$

The general solution of a linear differential equation consists of the sum two terms (see also Problem 3.3). The first one is the solution of the *homogeneous* equation (3.12), which contains an arbitrary constant. The second term is a *particular* solution of the inhomogeneous equation (3.11) and contains no degrees of freedom.

If the function g in (3.11) is constant and the *forcing term* h in (3.11) has a nice form, the above observation yields an easier alternative to the use of variation of constants: if we are able to guess a particular solution of the linear differential equation, we can simply add the solution of the homogeneous equation to obtain the general solution.

Example 3.2 Consider the linear non-autonomous differential equation

$$\frac{dx}{dt} = -x + t^2.$$

The homogeneous solution x_h is given by

$$x_h(t) = x_h(0)e^{-t},$$

where $x_\mathrm{h}(0)$ is an arbitrary constant. Since the forcing term is a second-order polynomial, we can also use a second-order polynomial as the *Ansatz* for our particular solution x_p, i.e.

$$x_\mathrm{p}(t) = at^2 + bt + c.$$

Substituting this ansatz into the differential equation gives

$$2at + b = (1-a)t^2 - bt - c,$$

and because the above has to hold for all t, we can equate equal powers of t to arrive at

$$a = 1, \qquad b = -2a = -2, \qquad c = -b = 2.$$

It follows that the general solution of the differential equation is given by

$$x(t) = (x(0) - 2)e^{-t} + t^2 - 2t + 2.$$

* * *

3.2.3.3 Solving First-Order Autonomous Linear Differential Equations

For autonomous differential equations we try to find a function x(t) whose derivative dx/dt is equal to $f(x)$. Autonomous scalar differential equations can be solved by a technique called *separation of variables*. The basic idea is to treat the left-hand side of (3.4) as a fraction, such that we can multiply (3.4) by $\mathrm{d}t$ and therefore separate the dependent and independent variable, resulting in

$$\frac{1}{f(x)}\,\mathrm{d}x = \mathrm{d}t. \tag{3.17}$$

Integrating this separated form then results in

$$\int_{x(0)}^{x(t)} \frac{1}{f(u)}\,\mathrm{d}u = \int_0^t \mathrm{d}s = t, \tag{3.18}$$

where we have used the dummy variables u and s to avoid confusion, and have used $t = 0$ as a convenient but arbitrary starting point. If we are able to solve the implicit equation (3.18), we can obtain an explicit solution $x(t)$ which, as expected, depends on the initial value $x(0)$.

Example 3.3 Consider the autonomous linear differential equation

$$\frac{\mathrm{d}x}{\mathrm{d}t} = ax,$$

where a is a given constant. You might already be familiar with this equation and know that it corresponds to exponential growth (or decay). Indeed, separating and integrating results in

$$\ln(x(t)) - \ln(x(0)) = at.$$

Taking the exponential at both sides of the above equation then yields

$$x(t) = x(0)e^{at}.$$

with $x(0)$ a constant defined by the initial value.

* * *

3.3 Geometric Reasoning, Equilibria and Stability

Let us return to the general expression for an autonomous differential equation, (3.4). We now wish to know the change of x (the movement) in a particular region on the x-axis from a starting position, i.e. an *initial value*. We prescribe x at time 0 to be x_0:

$$\begin{cases} \dfrac{dx}{dt} = f(x) \\ x(0) = x_0. \end{cases} \tag{3.19}$$

We will denote the solution of (3.19) by $x(t; x_0)$ and call this an *orbit*. An orbit, therefore, is a collection of points related by the evolution function of the dynamical system.

It is often helpful to study these equations from a geometric viewpoint by interpreting them as *vector fields*. In this way, we can determine the *qualitative behavior of solutions without having to solve the differential equation explicitly*. In fact, most nonlinear differential equations cannot be solved analytically. However, a global understanding of the behaviour of a system described by a nonlinear differential equation can be obtained by studying its dynamics in a vector field. For example, assume that our first-order nonlinear differential equation is given by

$$\dot{x} = f(x) = (x+2)(x+1)(x-1)^2 \tag{3.20}$$

and we wish to know for which values of x the system is in equilibrium, $\dot{x} = 0$, and if the equilibrium is stable or unstable. An equilibrium is *stable* if a small change away from the position will not (after waiting sufficiently long) move the system away from this equilibrium. If the equilibrium is *unstable*, small perturbations will move the system away from the equilibrium. A simple example is a marble in a bowl. In the bottom of the bowl, the equilibrium is stable as a small perturbation will return the marble to its previous (stable) position. This can be contrasted with a marble on top of a bowl: any small perturbation will result in the marble moving

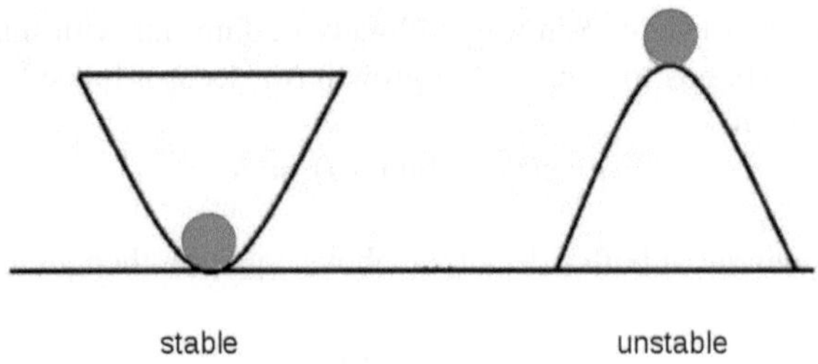

Fig. 3.2 Illustration of a stable and unstable equilibrium

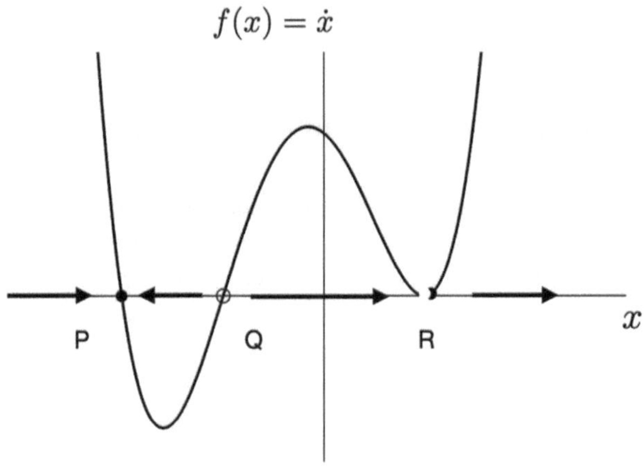

Fig. 3.3 Vector field for $f(x) = \dot{x} = (x + 2)(x + 1)(x - 1)^2$. The fixed points are shown, where a stable fixed point is marked with a closed circle, an unstable fixed point with an open circle. The changes in x are indicated with the arrows. Near a stable fixed point, the arrows point towards it; if the fixed point is unstable, the direction of the arrows is away from it. The point P is stable, Q is unstable and R is semi-stable. If we start at a point $P < x_0 < Q$, the orbit will start moving to the left because the function $f(x) < 0$ in this region, therefore x will decrease. The orbit cannot pass P, as the velocity is zero at this point. In fact this orbit never arrives (in finite time) exactly at P. If $x < P$ it holds that $f(x) > 0$ and x will move to the right. Similar arguments apply to Q and R

away from this position (Fig. 3.2). Returning to our (3.20), it is easy to determine the equilibria: $x = -2$, $x = -1$ and $x = 1$. These points are called **fixed points**. We now wish to know if a particular equilibrium is stable or unstable. Recall that the equilibrium is stable if x will evolve towards it and unstable it will move away from the equilibrium. The equilibria that are (un-)stable are known as an *(un-)stable fixed points*. We can determine if a fixed point is stable or unstable by simple graphical inspection of the function $f(x)$, shown in Fig. 3.3. If $f(x)$ is positive, the change in x, $\frac{dx}{dt} > 0$ and therefore x increases. If $f(x)$ is negative, the change is negative and "movement" is to the left, i.e. x decreases. By plotting this direction with arrows on the horizontal axis, the stability of the fixed points is directly clear. The equilibrium Q is unstable: after a small perturbation from Q the orbit will either go to P or R.

Such a perturbation can for instance be the effect of noise in the system, something that we do not explicitly consider here. On the other hand, a small perturbation from P is without consequences, as the orbit will return to P, in fact exponentially fast: P is a stable fixed point. For obvious reasons, the equilibrium R is *semi-stable*. A more extensive (and more formal) discussion of stability, including marginal and exponential stability is outside our scope.

3.4 Stability Analysis

As an alternative to the graphical techniques we discussed to determine the stability of a fixed point it is also possible to determine the stability by linearizing around a fixed point. This will provide both information about the nature of the fixed point (stable or unstable equilibrium) and the rate of change towards or from the fixed point. It can be shown (cf. Fig. 3.3) that the slope of the derivative of $f(x)$ evaluated at the fixed point determines stability: if $f'(x^*) > 0$ the fixed point is unstable and if $f'(x^*) < 0$ the fixed point is stable. If you look back at previous figures you can check that this is indeed the case. To determine the stability properties of an equilibrium, and assuming that the function f is continuously differentiable, we can use the following theorem.

Theorem 3.1 *Let f be a continuously differentiable scalar vector field.*

1. *If $f(x^*) = 0$ and $f'(x^*) < 0$, then x^* is an asymptotically stable equilibrium.*
2. *If $f(x^*) = 0$ and $f'(x^*) > 0$, then x^* is an unstable equilibrium.*

Compare this with the graph presented in Fig. 3.3. Fixed points determine the dynamics of first-order autonomous systems. Orbits either increase or decrease monotonically, or remain constant at the equilibria. This implies that oscillations in or periodic solutions to systems characterized by $\dot{x} = f(x)$ are impossible.

3.5 Bifurcations

In first order autonomous systems the system will always move either towards a stable equilibrium or depart (forever) from an unstable equilibrium, depending on the characteristics of the fixed points and our initial conditions. However, many autonomous dynamical systems may also depend on a particular parameter. For example, consider

$$\dot{x} = f(x, a) = x(x - a) \tag{3.21}$$

with a our parameter. The fixed points of (3.21) are $x = 0$ and $x = a$. Using our graphical analysis, we can derive that if $a > 0$ and $x > a$ then $\dot{x} > 0$ and if $0 < x < a$ then $\dot{x} < 0$, and therefore a is an unstable fixed point. For values $x < 0, \dot{x} > 0$, which implies that $x = 0$ is stable. For values $a < 0$ the situation is reversed, i.e. a is a stable fixed point and $x = 0$ is unstable. The fixed point $x = 0$ changes, therefore, from a stable fixed point to an unstable fixed point as the parameter a changes from $a > 0$ to $a < 0$, and this happens at exactly the value $a = 0$. This change in stability of a fixed point is an example of a *bifurcation*. Bifurcations can also result in the disappearance of equilibria, the appearance of a new equilibrium or a change from an unstable to an stable equilibrium. A bifurcation occurs, therefore, if there is a change in the nature of an equilibrium or if fixed points appear or disappear resulting from a change in the value of a particular parameter that describes the system.

Such changes in equilibrium will generally result in qualitative changes in the dynamics of the system. As a clinical example, a bifurcation may occur in the transition from normal brain function to the start of seizure or a psychosis. Identification of the relevant control parameter (or parameters) may provide a means to prevent this event from happening with obvious clinical relevance.

Bifurcations are further classified by the particular change in equilibria. If fixed points are *created* or *destroyed*, the bifurcation is a *saddle node bifurcation*. If equilibria change from stable to unstable, a *transcritical bifurcation* occurs. In a *pitchfork bifurcation* fixed points appear or disappear in symmetrical pairs. The codimension of a bifurcation is the number of parameters which must be varied for the bifurcation to occur. In our example, with the single parameter a, the bifurcation is codimension-one. We will now discuss these three bifurcations in more detail.

3.5.1 Saddle Node Bifurcation

Let us illustrate the saddle node bifurcation in a graphical manner starting with the vector field from Fig. 3.3. We add a parameter a that can move the curve upwards or downwards, and we therefore write $f(x, a)$. When the curve moves upwards, the equilibrium R will disappear. When it moves down, the equilibrium R splits into two equilibria, one being stable and the other being unstable. See Fig. 3.4. We also observe that the fixed points at P and Q are slightly perturbed by moving the graph of f up and down, but their stability properties remain unchanged. The red curve corresponds to the so called *critical case*. At R it now holds that the derivative is equal to zero, that is it corresponds to the quadratic tangency of f at this point. For this equilibrium to exist we have therefore to evaluate the *partial derivative* of

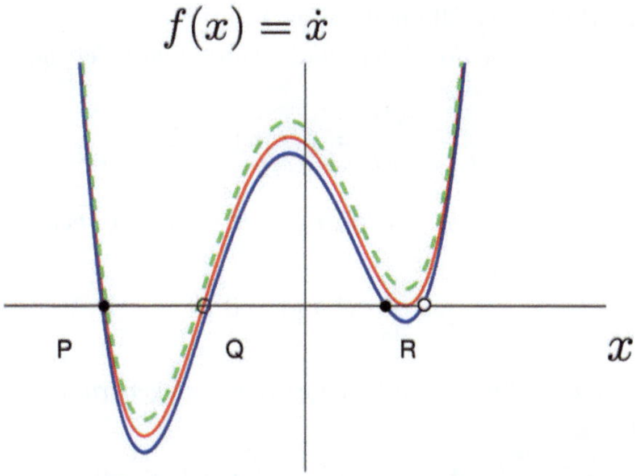

Fig. 3.4 The solid red line is the graph of f as in Fig. 3.3. The dashed green line corresponds to the graph of $f + 0.2$. Note that the fixed point at R has disappeared. The dashed blue line corresponds to the graph of $f - 0.2$: the fixed point at R has split into two fixed points. The flow in between these fixed points is to the left, and consequently the left of these two fixed points is asymptotically stable (filled circle), while the right of these two fixed points is an unstable point (open circle). While the fixed points P and Q have slightly changed their position, their stability remains unchanged

the function f with respect to x at the fixed point $x^* = R$, and this should equal zero: $\partial f(x^*, a)/\partial x = 0$. You can easily argue why this is the case: as a bifurcation can occur if an equilibrium changes from stable to unstable (or vice verse), and we learned already in Theorem 3.1 that the sign of $f'(x^*)$ defines if an equilibrium is stable ($f'(x^*) < 0$) or unstable ($f'(x^*) > 0$), a necessary condition for a change in stability is that $f'(x^*, a) = 0$.

For a saddle node bifurcation to occur, we need two additional conditions.[4] First, the second partial derivative of $f(x, a)$ with respect to x at R is must be nonzero, and the first partial derivative of $f(x, a)$ with respect to the control parameter, a, needs to be nonzero.[5] If all these three conditions are satisfied, a fixed point is *created* or *destroyed* and a *saddle node bifurcation* occurs. These conditions are summarized in Theorem 3.2.

[4]These two conditions are known as non-degeneracy conditions. For instance, if the second derivative would be zero, then a saddle node bifurcation is not guaranteed. An example is given by the equation $\dot{x} = x^3 + a$.

[5]It means that there is a truly quadratic tangency at the fixed point and the eigenvalue is able to cross $f(x, a) = 0$ as the control parameter is changed.

Theorem 3.2 (Fold or saddle-node bifurcation)
Suppose that a smooth scalar differential equation, that depends on a parameter a,

$$\frac{dx}{dt} = f(x, a)$$

has an equilibrium $x = x^*$ at $a = a_0$, i.e. $f(x^*, a_0) = 0$, such that

$$\frac{\partial f}{\partial x}(x^*, a_0) = 0$$

and assume that in addition the following two non-degeneracy (ND) conditions are satisfied:

$$\frac{\partial^2 f}{\partial x^2}(x^*, a_0) \neq 0; \qquad \frac{\partial f}{\partial a}(x^*, a_0) \neq 0$$

then a fold or saddle-node bifurcation is present at $f(x^*, a_0)$.

Example 3.4 An example for a saddle node bifurcation is given by the first order autonomous nonlinear differential equation

$$\dot{x} = x^2 + r \tag{3.22}$$

where the parameter r defines the number of fixed points. If the parameter $r < 0$ a stable and an unstable equilibrium exist, that merge if $r = 0$ into a half-stable fixed point at $x = 0$. This equilibrium vanishes if $r > 0$, leaving no fixed points. The saddle node bifurcation occurred at $r = 0$ since here the equilibrium vanished and the resulting vector fields are now qualitatively different for negative and positive values of r. This is graphically illustrated in Fig. 3.5. If we apply Theorem 3.2 to our (3.22) and consider the first and second partial derivatives of the function $f(x, r) = x^2 + r$ we can indeed show that the equilibrium at $x = 0$ for $r_0 = 0$ is at a saddle-node bifurcation: $\frac{\partial f}{\partial x}(x^*, r_0) = 0$ and both non-degeneracy conditions are satisfied as $\frac{\partial^2 f}{\partial x^2}(x^*, r_0) \neq 0$ and $\frac{\partial f}{\partial r}(x^*, r_0) = 1 \neq 0$.

We can also draw a graphical representation of the relation between the control parameter r and the presence of stable or instable equilibria for values of x. Such representation is called a *bifurcation diagram*. The bifurcation diagram of (3.22) is shown in Fig. 3.6. Note, that if we start with $r > 0$, and r is decreased, the bifurcation results in the creation of a stable and an unstable equilibrium, while if we start with $r < 0$ and r is increased, the stable and unstable equilibrium disappear and no equilibrium exists if $r > 0$.

Each vertical slice of the bifurcation diagram depicts a phase space of the system for a particular parameter value. For example, for $r < 0$ in the diagram above, there are two equilibrium points, one stable (solid line) and the other unstable (dashed

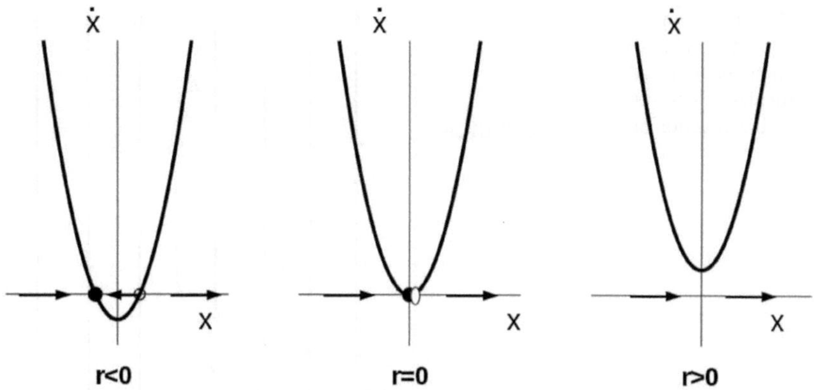

Fig. 3.5 Saddle-node bifurcation. Left: $r < 0$ results in a stable (open circle) and an unstable (closed circle) equilibrium. If $r = 0$, the two equilibria merge into a half-stable fixed point. For $r > 0$ no equilibrium exists. The bifurcation occurs at $r = 0$. Note that at the bifurcation it holds that the tangent is horizontal. As was summarized in Theorem 3.2, additional conditions need to be satisfied for a saddle-node bifurcation to occur

Fig. 3.6 Bifurcation diagram for the saddle-node bifurcation, defined by $\dot{x} = x^2 + r$. For values $r > 0$ no equilibria exist, for values $r < 0$ a stable (solid line) and unstable (dashed line) equilibrium exist at the values of x indicated

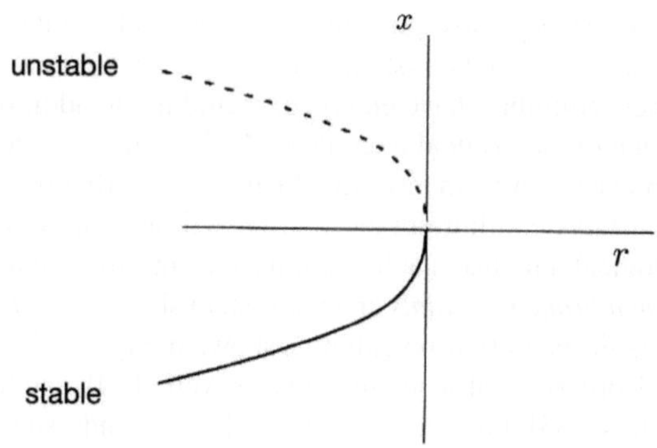

line). We can further visualize the ows of the systems state by drawing arrows. An upward arrow above the unstable equilibrium, a downward arrow below to unstable and above the stable equilibrium and an upward arrow below the stable one. Doing this for several values results in Fig. 3.7.

* * *

In neuroscience, saddle-node bifurcations can occur in various conditions, for instance from a change in synaptic currents, the availability of ATP or ion concentrations in the extracellular space. Creation of a new equilibrium resulting from a saddle node bifurcation, with preservation of an existing equilibrium, can also generate a new, additional, stable state. Returning to Fig. 3.4, assume we are at the red curve. There is only one stable equilibrium, close to the point P. This equilibrium could

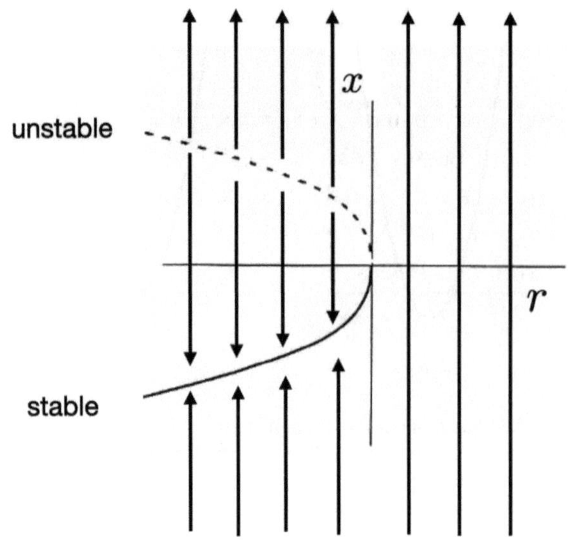

Fig. 3.7 Bifurcation diagram for the saddle-node bifurcation, including arrows for particular values of r showing the direction of the change in x

for instance correspond to a physiological membrane potential at rest of −60 mV. If parameters of the systems are changed such that we arrive at the solid blue curve, two new equilibria have emerged, including an additional *stable* equilibrium at the left side of the original fixed point R. This may e.g. correspond to a pathological membrane potential in rest, say -20 mV. A perturbation of the healthy state that crosses the unstable equilibrium close to Q, will now drive the equilibrium to this new pathological, but also stable, equilibrium. In this example, therefore, we have *two stable membrane potentials* and the system shows *bistability*. A real-world example of two stable membrane potentials is shown in Fig. 3.8. Bistability is present in some seizure disorders, as absence seizures, as well [110], and has been observed after metabolic stress [35]. Further, bistability is key for understanding basic phenomena of cellular functioning, such as decision-making processes in cell cycle progression, cellular differentiation and apoptosis. Bistability can also involved in the loss of cellular homeostasis associated with early events in cancer onset or some prion diseases.

3.5.2 *Transcritical Bifurcation*

In a transcritical bifurcation fixed points are not created or destroyed (as in a saddle node bifurcation) but the equilibria of fixed points exchange their stability when they collide: the stable fixed point becomes unstable and the unstable fixed point becomes stable. We illustrate this with the dynamical system given by

$$\dot{x} = f(x) = rx - x^2 \tag{3.23}$$

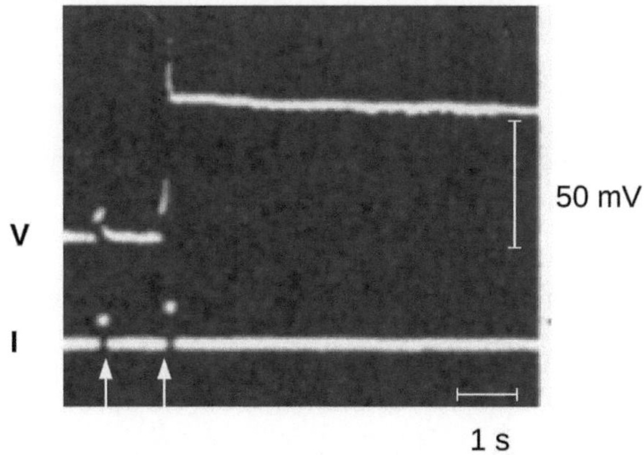

Fig. 3.8 Two stable membrane resting potentials in a myelinated toad axon in a potassium rich solution, recorded at the nodes of Ranvier. The top trace shows the membrane potential, the bottom trace the stimulation current. The nodes were hyperpolarized by a constant external source of current, I, at the critical intensity. Stimulation with a brief, small additional current (first arrow) does not result in a change in membrane potential. A slightly larger pulse, however, does cause the axon to go to a new stable membrane potential, approximately 50 mV above the other stable membrane potential. In this experiment, the extracellular potassium was 30 m-equiv/l, sodium 80 m-equiv/l and chloride 0 m-equiv/l. Adapted from Fig. 10-B in Tasaki, J Physiol 1959 [114]. Reproduced with permission from John Wiley and Sons

which looks like the logistic equation (but both x and r can take both negative and positive values). It is easy to verify that fixed points exist for values $x^* = 0$ and $x^* = r$ by setting $f(x) = 0$.

In Fig. 3.9 we plot f(x) versus x. For $r < 0$ (panel (a)) an unstable equilibrium exists at $x = r$ and stable fixed point at $x = 0$. Similar to what we observed at a saddle node bifurcation, if $r = 0$ the equilibria merge into a half-stable fixed point at $x = 0$ (panel (b)). If $r > 0$ the equilibrium $x = 0$ does not vanish, but becomes unstable while the unstable equilibrium at $x = r$ becomes stable (panel (c)). We therefore observe that if $r < 0$ then $x = 0$ is stable and $x = r$ is unstable; for $r = 0$, $x = 0$ is unstable and if $r > 0$ then $x = 0$ is unstable and $x = r$ is stable. So, as r passes through $r = 0$, the two steady states cross each other and exchange stability: a transcritical bifurcation. This results in the bifurcation diagram of Fig. 3.10). Similar to the Theorem for a fold or saddle-node bifurcation, we write the Theorem for the existence of a transcritical bifurcation for a fixed point.

Theorem 3.3 (Transcritical bifurcation) *Suppose that a smooth scalar differential equation, that depends on a parameter a,*

$$\frac{dx}{dt} = f(x, a)$$

has an equilibrium $x = x^*$ *at* $a = a_0$, *i.e.* $f(x^*, a_0) = 0$, *such that*

$$\frac{\partial f}{\partial x}(x^*, a_0) = 0; \qquad \frac{\partial f}{\partial a}(x^*, a_0) = 0$$

and assume that in addition the following two non-degeneracy conditions are satisfied

$$\frac{\partial^2 f}{\partial x^2}(x^*, a_0) \neq 0; \qquad \frac{\partial^2 f}{\partial x \partial a}(x^*, a_0) \neq 0$$

then a transcritical bifurcation is present at $f(x^*, a_0)$.

You can check that (3.23) indeed satisfies Theorem 3.3.

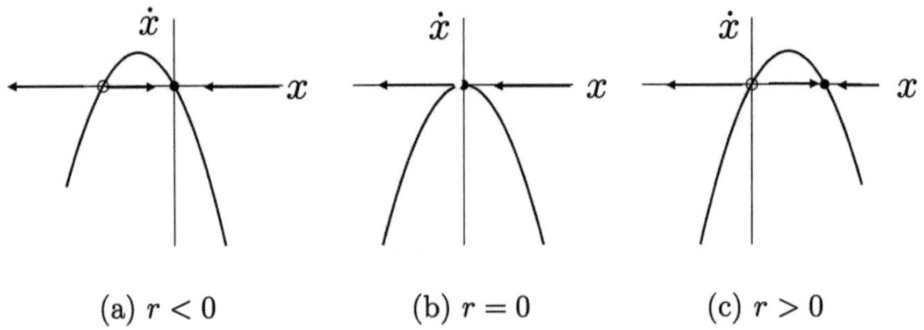

Fig. 3.9 Transcritical bifurcation, (3.23). Left panels show $\dot{x} = rx - x^2$ versus x for $r < 0, r = 0$ and $r > 0$. Solid circles: stable fixed points; open circles: unstable fixed points. The middle panel shows a half stable fixed point at the origin

Fig. 3.10 Bifurcation diagram of (3.23), with a transcritical bifurcation at the origin

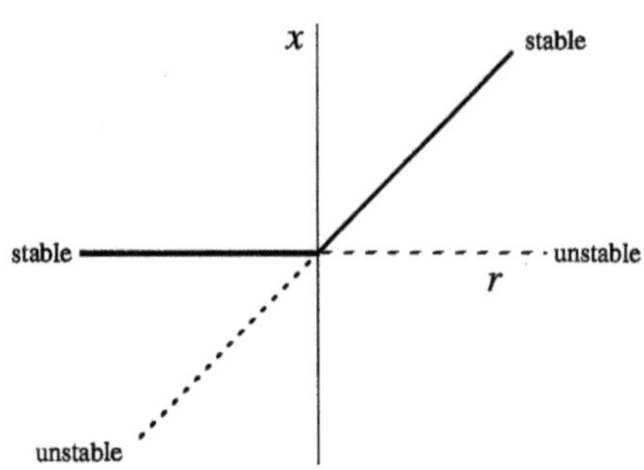

3.5.3 *Pitchfork Bifurcation*

In some systems fixed points appear or disappear in symmetrical pairs. An example from physics is the bending of a vertical beam with a load on top. If the load increases a bifurcation will occur where the beam will bend to either the left or the right side. At this critical value of the load, therefore, two new stable fixed points arise from an initial single stable fixed point and the previous position of the beam has now become unstable.

For pitchfork bifurcations, both appearance and disappearance of symmetrical fixed points occur. If a pair of stable fixed points appear, with initially small values compared to the (single) fixed points, the bifurcation is supercritical. An example is given by the equation $\dot{x} = rx - x^3$, shown in Fig. 3.11 for three different values of r. For values $r < 0$, $x = 0$ is a stable equilibrium and for $r > 0$ a pair of stable equilibria appears given by $x = \pm\sqrt{r}$ (and an unstable equilibrium $x = 0$), the bifurcation is known as a supercritical. The corresponding bifurcation diagram is shown in Fig. 3.12. The example of loading our vertical beam with a weight on top is a physical example of such a supercritical bifurcation.

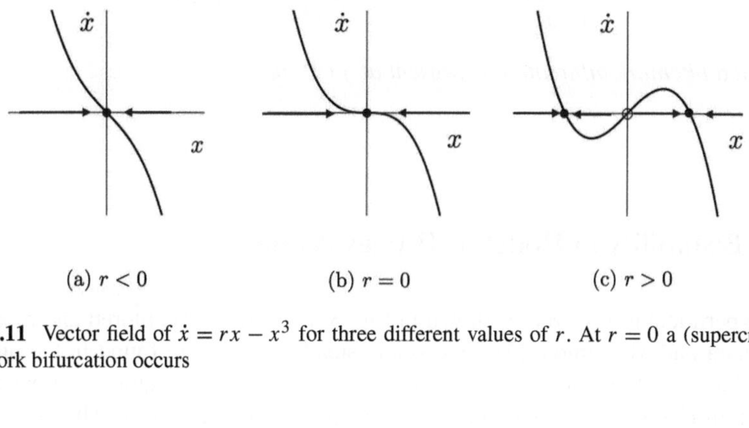

(a) $r < 0$ (b) $r = 0$ (c) $r > 0$

Fig. 3.11 Vector field of $\dot{x} = rx - x^3$ for three different values of r. At $r = 0$ a (supercritical) pitchfork bifurcation occurs

Fig. 3.12 Supercritical pitchfork bifurcation diagram. The control parameter r can be a measure for a weight if it applies to our loading of a vertical beam

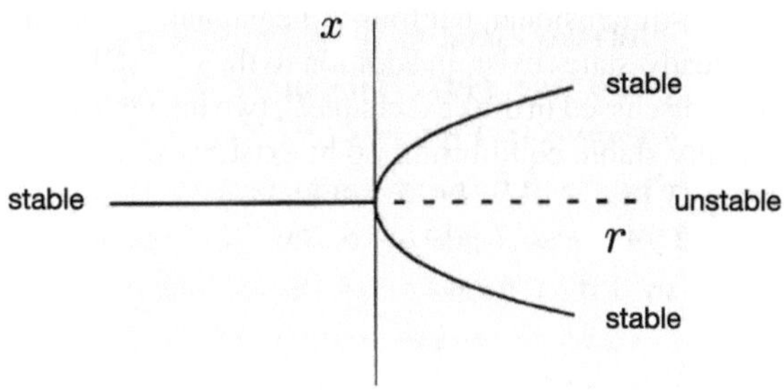

If we consider $\dot{x} = rx + x^3$, the nature of the bifurcation is different. For $r < 0$ there exists a pair of unstable fixed points ($x = \pm\sqrt{-r}$) and a stable fixed point ($x = 0$) exist that all disappear at $r = 0$ where for values $r > 0$ no stable fixed points remain: this is a subcritical bifurcation. For the pitchfork bifurcation, we also present the conditions that need to be satisfied for its occurrence in Theorem 3.4.

Theorem 3.4 (Pitchfork bifurcation) *Suppose that a smooth scalar differential equation, that depends on a parameter a,*

$$\frac{dx}{dt} = f(x, a)$$

has an equilibrium $x = x^*$ *at* $a = a_0$, *i.e.* $f(x^*, a_0) = 0$, *such that*

$$\frac{\partial f}{\partial x}(x^*, a_0) = 0; \qquad \frac{\partial f}{\partial a}(x^*, a_0) = 0; \qquad \frac{\partial^2 f}{\partial x^2}(x^*, a_0) = 0$$

and that in addition the two following two non-degeneracy conditions are satisfied

$$\frac{\partial^2 f}{\partial x \partial a}(x^*, a_0) \neq 0; \qquad \frac{\partial^3 f}{\partial x^3}(x^*, a_0) \neq 0$$

then a pitchfork bifurcation is present at $f(x^*, a_0)$.

3.6 Bistability in Hodgkin-Huxley Axons

The experimental observation shown in Fig. 3.8 was further explored and modeled by Aihara and Matsumoto [2]. Two stable steady states were found in the numerical solution of the Hodgkin-Huxley equations for the intact squid axon bathed in potassium-rich sea water with an externally applied inward current. The model was not one-dimensional, but four-dimensional. Under the conditions where the two stable steady-states exist, in addition to the two stable steady-states, a stable limit cycle (to be discussed in the next chapter), two unstable equilibrium points, and one asymptotically stable equilibrium point exist, as well. The bifurcation diagram[6] is shown in Fig. 3.13.

[6] Shown is a modified version; the limit cycle is not indicated.

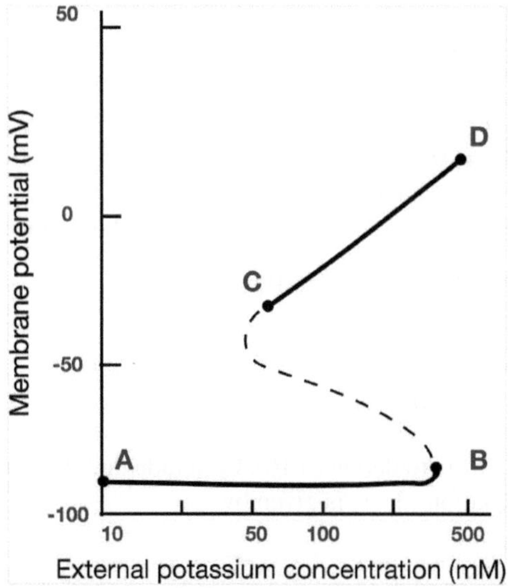

Fig. 3.13 Bifurcation diagram for the Hodgkin-Huxley equations. The transmembrane voltage is the bifurcation variable, while the external potassium concentration is the bifurcation parameter. Similar to the experiment illustrated in Fig. 3.8 a steady hyperpolarizing bias current of -20 μA/cm^2 is injected throughout. The S-shaped curve shows the set of steady membrane potentials. The solid curves between A and B and C and D represent the stable equilibrium points, the dashed curve represents the unstable equilibrium points. In the range of $[K]$ from 66.0 mM $< [K] <$ 417.0 mM (C-B),the Hodgkin-Huxley equations have two asymptotically stable equilibrium points Note the presence of hysteresis, as well, to be discussed in the next chapter. Adapted from Aihara and Matsumoto [2], with permission from The Biophysical Society, Elsevier Inc

3.7 Summary

You have learned about dynamics and why this is relevant in biological systems. We discussed that differential equations are essential tools to study dynamical systems, and that graphical tools can assist in understanding the global behavior of systems that can be described by differential equations without explicitly solving the equations. You understand what equilibria are and bifurcations. You also learned that oscillations are not observed for scalar differential equations $\dot{x} = f(x)$ on the line (but it can on a circle). In the next chapter, we will add a dimension ($n = 2$) and learn that in those situations oscillations are possible.

Problems

3.1 Solve the following initial value problems.

a. $\dot{x} = x^2, x(0) = 1$.
b. $\dot{x} = x^2 + 1, x(0) = 0$.
c. $\tau\dot{x} = -x, x(0) = 1$. Determine the time t where it holds that $x(t) = \frac{1}{2}$. The constant τ is called the *time constant*. Explain why this makes sense.

3.2 Solve the following initial value problems.

a. $\frac{dx}{dt} = -\cos(t)x, \qquad x(0) = 1.$
b. $\frac{dx}{dt} = \sin(t)x^2, \qquad x(0) = 1.$

3.3 Use (3.14) and (3.11) to derive (3.16). Conclude that the general solution of the linear differential equation (3.11) is given by

$$x(t) = x(0)e^{G(t)} + \int_0^t e^{G(t)-G(s)}h(s)\,ds.$$

3.4 Consider the periodically forced differential equation

$$\frac{dx}{dt} = -2x + \sin(\omega t),$$

where ω is the frequency of the forcing term.

a. Use the ansatz $x_p(t) = a\cos(\omega t) + b\sin(\omega t)$ to determine a particular solution of the above equation.
b. Determine the general solution of the differential equation.
c. Argue why, after transients, the general solution closely follows the particular solution.
d. Plot the maximal amplitude of the particular solution as a function of the forcing frequency omega.
e. Explain why this ODE models a low-pass filter.

3.5 Consider $\dot{x} = f(x)$ with $f(x) = -x$ and $x_0 = -1$. Compute the orbit explicitly. Show that

$$lim_{t\to\infty} x(t; -1) = 0.$$

3.6 Consider the differential equation

$$\dot{x} = \sin(x).$$

Where do fixed points occur? Are they stable or unstable?

3.7 Consider the equation

$$\dot{x} = x(1 - x)$$

with initial condition $x(0) = x_0$.

a. Sketch the fixed points and indicate their stability.
b. Find an explicit expression for x as a function of t.
c. In the situation where $x(t) \to -\infty$, show that this is reached in finite time and calculate how long this takes. This is called the 'escape time'.

3.8 Consider a population with size N. Its size is governed by the following equation

$$\dot{N} = rN\left(1 - \frac{N}{K}\right)$$

where the constant $r > 0$ defines the growth rate and K is the carrying capacity. This equation is known as the logistic equation and describes the growth of a population, proposed by Verhulst in 1838. If we start with a small population size N, the growth is initially mainly governed by rN. However, as N grows, the second term $rN^2K)$ becomes more relevant as competition will start within the population with size N for critical resources, such as food or living space. Here, the parameter K models the strength of this contribution. Although this equation can be solved analytically, we wish to study its qualitative behavior by a graphical approach. Sketch the vector field, i.e. the relation between $\dot{N}$ and N and indicate the equilibria. What is the final size of the population?

3.9 Draw the vector fields of the following systems, determine the equilibria and if they are stable or unstable.

a. $\dot{x} = x(1 - x)(x - a), \qquad a \in (0, 1)$
b. $\dot{x} = x - x^3$

3.10 Consider

$$\dot{x} = x(x - a).$$

Plot the function $f = x(x - a)$ for values of a in the range $[-1; 1]$. What happens with the stability of the origin at $a = 0$?

3.11 For each of the exercises, plot the vector fields for a few values of λ. Show that a saddle-node bifurcation occurs for (a) particular value(s) of λ (don't forget to check the non-degeneracy conditions). Sketch the bifurcation diagram of the fixed points versus λ.

a. $\dot{x} = -x + x^2 + \lambda$
b. $\dot{x} = -x + x^3 + \lambda$
c. $\dot{x} = x^2 + \lambda x + 1$
d. $\dot{x} = x^2 - \lambda$
e. $\dot{x} = x^2 + 2x + \lambda^2$

3.12 Tumor growth has been the topic of many modeling attempts, including the effects of treatment. A well-known model for tumor growth is given by

$$\dot{N} = -aN \ln(bN)$$

with $a, b > 0$ parameters and N the number of cancerous cells. This equation is known as Gompertz law, and many variations to this original equation have been studied as well.
(a) Can you interpret the parameters $a, b > 0$?
(b) Sketch the vector field and make a graph of N as a function of time for various initial values.

3.13 Consider

$$\dot{x} = x^3 - ax.$$

Show that three equilibria can exist, depending on the value of a.

Chapter 4
Dynamics in Two-Dimensional Systems

The mind is the music that neural networks play.

—T.J. Sejnowski

Abstract In this chapter we discuss two dimensional systems of differential equations. We treat fixed points and set rules to define the stability for linear and nonlinear equations. In the second part, we discuss bifurcations, and show various examples of emergence or disappearance of oscillatory behavior, including limit cycles, one of the key characteristics of neurons and neuronal populations.

4.1 Introduction

In the previous chapter we discussed differential equations (ODEs) of one dependent variable and observed that the orbits are either constant or move monotonically towards or from equilibria. In this chapter, we consider autonomous systems in two-dimensions. An important difference with the autonomous scalar (one-dimensional) equations is that solutions of autonomous planar (2D) differential equations can show a much richer dynamics, including *periodic solutions*. Such oscillations are abundantly present in biology, both in physiological and pathological situations. We treat the general aspects of systems of (two-dimensional) autonomous differential equations with various examples from biology and neuroscience. While most systems in biology are nonlinear, linear equations will be discussed first. We show how to determine the solutions for linear equations and the stability of fixed points. Thereafter, we will treat the nonlinear case, and explain that stability of fixed points for nonlinear systems can be estimated by linearization around the fixed point of interest, making it possible to use the techniques we learned for the linear situation.

Electronic supplementary material The online version of this chapter (https://doi.org/10.1007/978-3-662-61184-5_4) contains supplementary material, which is available to authorized users.

The original version of this chapter was revised: Figure 4.3 has been updated. The correction to this chapter is available at https://doi.org/10.1007/978-3-662-61184-5_11

M. J. A. M. van Putten, *Dynamics of Neural Networks*,
https://doi.org/10.1007/978-3-662-61184-5_4

4.2 Linear Autonomous Differential Equations in the Plane

The general form for an autonomous system of $n = 2$ first order linear equations with constant, real coefficients is

$$\begin{aligned} \dot{x} &= ax + by \\ \dot{y} &= cx + dy \end{aligned} \tag{4.1}$$

which can also be written in a more compact form as

$$\dot{\boldsymbol{x}} = \boldsymbol{A}\boldsymbol{x} \tag{4.2}$$

with $\boldsymbol{x} = \begin{pmatrix} x \\ y \end{pmatrix}$ and coefficient matrix $A = \begin{pmatrix} a & b \\ c & d \end{pmatrix}$. We know that the general solution of (4.2) is equivalent to a second order homogeneous linear differential equation with a general solution that contains two linearly independent parts.[1] The solution will consist of some type of exponential functions. Let us assume that (4.2) has a solution of the form

$$\boldsymbol{\xi} e^{\lambda t} = \begin{pmatrix} \xi_1 \\ \xi_2 \end{pmatrix} e^{\lambda t}. \tag{4.3}$$

If we substitute this into (4.1), we obtain

$$\lambda \boldsymbol{\xi} e^{\lambda t} = \mathbf{A} \boldsymbol{\xi} e^{\lambda t}. \tag{4.4}$$

Using $e^{\lambda t} \neq 0$, we obtain

$$\boldsymbol{A}\boldsymbol{\xi} = \lambda \boldsymbol{\xi}. \tag{4.5}$$

This implies that $\boldsymbol{\xi} e^{\lambda t}$ is a solution of (4.1) if $\boldsymbol{\xi}$ is an *eigenvector* and λ is an *eigenvalue* of the matrix $\boldsymbol{A}$. To obtain the eigenvectors and eigenvalues, we proceed with (4.5), using $\boldsymbol{I} = \begin{pmatrix} 1 & 0 \\ 0 & 1 \end{pmatrix}$ to arrive at

$$(\boldsymbol{A} - \lambda \boldsymbol{I})\boldsymbol{\xi} = \begin{pmatrix} 0 \\ 0 \end{pmatrix}. \tag{4.6}$$

The eigenvalues of $\boldsymbol{A}$ are now given by the characteristic equation $\det(\boldsymbol{A} - \lambda \boldsymbol{I}) = 0$. This results in

$$\det \begin{pmatrix} a - \lambda & b \\ c & d - \lambda \end{pmatrix} = 0 \tag{4.7}$$

[1] Every nth order linear differential equation is equivalent to a system of n first order linear equations. As an example, for $n = 2$, take $a_2\ddot{y} + a_1\dot{y} + a_0 y = 0$ and set $y = x_1$ and $\dot{y} = x_2$. This results in an equivalent *system* of two first order linear equations $\dot{x_1} = x_2$ and $\dot{x_2} = -\frac{a_0}{a_2}x_1 - \frac{a_1}{a_2}x_2$.

or

$$\lambda^2 - \tau\lambda + \Delta = 0 \tag{4.8}$$

with solution

$$\lambda_i = \frac{\tau \pm \sqrt{\tau^2 - 4\Delta}}{2} \tag{4.9}$$

with $i = 1, 2$ and

$$\begin{aligned} \tau &= \text{trace}(\boldsymbol{A}) = a + d \\ \Delta &= \det(\boldsymbol{A}) = ad - bc. \end{aligned} \tag{4.10}$$

We subsequently solve, for each eigenvalue,

$$\begin{pmatrix} a - \lambda & b \\ c & d - \lambda \end{pmatrix} \boldsymbol{\xi} = \begin{pmatrix} 0 \\ 0 \end{pmatrix} \tag{4.11}$$

to obtain the eigenvectors $\boldsymbol{\xi}$. Three possibilities exist. The matrix $\boldsymbol{A}$ has (i) two distinct real eigenvalues, (ii) it has complex conjugate eigenvalues and (iii) it has a repeated eigenvalue. We will discuss these three cases in more detail.

4.2.1 *Case 1: Two Distinct Real Eigenvalues*

If the coefficient matrix $\boldsymbol{A}$ has two distinct real eigenvalues, the general solution of (4.2) is

$$\boldsymbol{x}(t) = c_1 \boldsymbol{\xi}_1 e^{\lambda_1 t} + c_2 \boldsymbol{\xi}_2 e^{\lambda_2 t}. \tag{4.12}$$

This is a linear combination of solutions, and it satisfies the initial conditions. By the existence and uniqueness theorem, it can then be proven that this is the only solution. The evolution of the system $\boldsymbol{x}(t)$ is thus according to a linear combination of changes in the direction of the two eigenvectors, where the velocity of the change in the direction of each eigenvector is defined by the corresponding eigenvalue.

Example 4.1 Solve

$$\begin{aligned} \dot{x} &= -2x + 2y \\ \dot{y} &= x - y. \end{aligned} \tag{4.13}$$

with $\boldsymbol{x}(0) = (5, -1)$. The coefficient matrix $A = \begin{pmatrix} -2 & 2 \\ 1 & -1 \end{pmatrix}$ with $\Delta = 0$ and $\tau = -3$. The eigenvalues follow from (4.9): $\lambda_1 = 0$ and $\lambda_2 = -3$. The corresponding eigenvectors are $\boldsymbol{\xi}_1 = (1, 1)^T$ and $\boldsymbol{\xi}_2 = (2, -1)^T$. The solution of (4.13) is therefore

$$\boldsymbol{x}(t) = \begin{pmatrix} 1 \\ 1 \end{pmatrix} + 2 \begin{pmatrix} 2 \\ -1 \end{pmatrix} e^{-3t}. \tag{4.14}$$

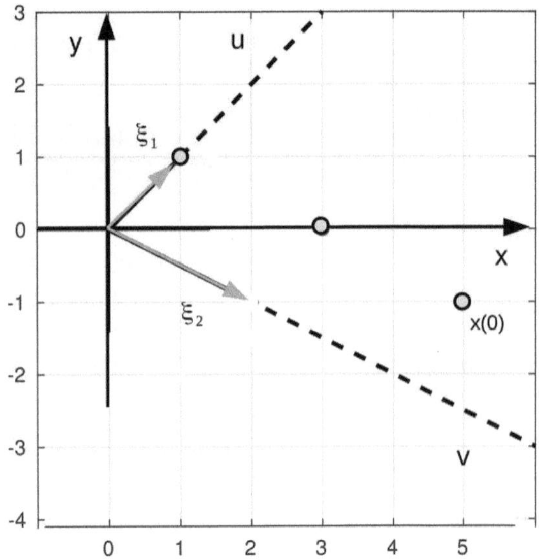

Fig. 4.1 Shown are two coordinate systems. First, the x-y plane, with the orthogonal x and y-axis. Second, the u-v plane, where the basis is formed by the eigenvectors $\boldsymbol{\xi}_1$ and $\boldsymbol{\xi}_2$, forming an 'eigenbasis'. Starting at initial condition $\boldsymbol{x}(0) = (5, -1)$ we have drawn $\boldsymbol{x}(t > 0)$ and $\boldsymbol{x}(t \to \infty) = (1, 1)$. In this example, the change is along the direction of the eigenvector ξ_2, the direction of v. In the u-v plane the initial condition has coordinates $(u, v) = (1, 2)$

Let's discuss this in some more detail, as it may assist in understanding the relation of eigenvectors and eigenvalues in the solution of linear 2D systems. We draw a few solutions in the x-y plane starting at $\boldsymbol{x}(0) = (5, -1)$, shown in Fig. 4.1. We can also express the initial condition in the u-v plane where the basis is formed by the eigenvectors ξ_1 and ξ_2. This alternative basis is called an *eigenbasis*. Note, that in the eigenbase coordinate system, the initial condition is $(u, v) = (1, 2)$. The solution moves in the direction of the eigenvectors as time evolves, where the velocity is defined by the eigenvalues.

Example 4.2 In Chap. 2 we discussed the interaction of a neurotransmitter with a ligand-gated channel, and obtained (2.6). We now rewrite this in matrix notation

$$\dot{\boldsymbol{x}} = \begin{pmatrix} -\beta - k_{-1} & \alpha \\ \beta & -\alpha \end{pmatrix} \boldsymbol{x} \tag{4.15}$$

where $\boldsymbol{x} = (x, y)^T$, with x the AR-complex and y the AR_{open} complex. We assume that a solution of the form $\boldsymbol{\xi} e^{\lambda t}$ exists. For the trace we find $\tau = -\beta - k_{-1} - \alpha$ and for the determinant $\Delta = -\alpha(-\beta - k_{-1}) - \alpha\beta = \alpha k_{-1}$. The eigenvalues are $\lambda_1 = 0.5(\tau + \sqrt{\tau^2 - 4\Delta})$ and $\lambda_2 = 0.5(\tau - \sqrt{\tau^2 - 4\Delta})$, both real and distinct. We now solve for each eigenvalue

$$\begin{pmatrix} -\beta - k_{-1} - \lambda_i & \alpha \\ \beta & -\alpha - \lambda_i \end{pmatrix} \boldsymbol{\xi} = \begin{pmatrix} 0 \\ 0 \end{pmatrix} \tag{4.16}$$

to obtain the eigenvectors

$$\xi_1 = \begin{pmatrix} \frac{\alpha+\lambda_1}{\beta} \\ 1 \end{pmatrix}, \; \xi_2 = \begin{pmatrix} \frac{\alpha+\lambda_2}{\beta} \\ 1 \end{pmatrix}. \tag{4.17}$$

The general solution is therefore

$$\boldsymbol{x} = c_1 \begin{pmatrix} \frac{\alpha+\lambda_1}{\beta} \\ 1 \end{pmatrix} e^{\lambda_1 t} + c_2 \begin{pmatrix} \frac{\alpha+\lambda_2}{\beta} \\ 1 \end{pmatrix} e^{\lambda_2 t}. \tag{4.18}$$

With initial conditions $x(0) = AR(0)$ and $y(0) = 0$ we obtain for the number of open channels AR_{open}

$$y(t) = AR_{\text{open}}(t) = c\left(e^{\lambda_1 t} - e^{\lambda_2 t}\right) = \frac{AR(0)\beta}{\lambda_1 - \lambda_2}\left(e^{\lambda_1 t} - e^{\lambda_2 t}\right) \tag{4.19}$$

which is our (2.7) from Chap. 2 describing the time-dependent synaptic conductance.

4.2.2 *Case 2: Complex Conjugate Eigenvalues*

In the two previous examples, both eigenvalues were real and distinct. Let us now take our homogeneous system with constant coefficients (4.2), here repeated for convenience,

$$\dot{\boldsymbol{x}} = A\boldsymbol{x}$$

with $A = \begin{pmatrix} 1 & -1 \\ 5 & -3 \end{pmatrix}$ and evaluate the characteristics of the fixed point. We find that our eigenvalues will be $\lambda_1 = -1 + i$ and $\lambda_2 = -1 - i$, with eigenvectors $\boldsymbol{\xi}_1 = (1, \; 2 - i)$ and $\boldsymbol{\xi}_2 = (1, \; 2 + i)$. In this case, a fundamental set of solution is

$$\boldsymbol{x}_1 = \begin{pmatrix} 1 \\ 2 - i \end{pmatrix} e^{(-1+i)t}, \; \boldsymbol{x}_2 = \begin{pmatrix} 1 \\ 2 + i \end{pmatrix} e^{(-1-i)t} \tag{4.20}$$

A real-valued solution is now obtained by taking the real and imaginary part of either $\boldsymbol{x}_1$ or $\boldsymbol{x}_2$ (we state this without proof, here). This results in

$$\boldsymbol{x} = c_1 e^{-t} \begin{pmatrix} \cos t \\ 2\cos t + \sin t \end{pmatrix} + c_2 e^{-t} \begin{pmatrix} \sin t \\ -\cos t + 2\sin t \end{pmatrix} \tag{4.21}$$

The corresponding fixed point for this system is a stable spiral: for $t \to \infty$ the system will evolve to the origin.

The general solution of (4.2) in the case of complex conjugate eigenvalues $\lambda = \rho \pm i\mu$ and eigenvector $\boldsymbol{\xi}_1 = \boldsymbol{a} + i\boldsymbol{b}$ and $\boldsymbol{\xi}_2 = \boldsymbol{a} - i\boldsymbol{b}$, where $\boldsymbol{a}$ and $\boldsymbol{b}$ are real, is given by

$$\boldsymbol{x} = c_1\boldsymbol{u}(t) + c_2\boldsymbol{v}(t) \tag{4.22}$$

with

$$\begin{aligned} \boldsymbol{u}(t) &= e^{\rho t}(\boldsymbol{a}\cos\mu t - \boldsymbol{b}\sin\mu t), \\ \boldsymbol{v}(t) &= e^{\rho t}(\boldsymbol{a}\sin\mu t - \boldsymbol{b}\cos\mu t). \end{aligned} \tag{4.23}$$

4.2.3 *Case 3: Repeated Eigenvalue*

We now consider the scenario where, in solving (4.2),

$$\dot{\boldsymbol{x}} = \boldsymbol{A}\boldsymbol{x}$$

the matrix $\boldsymbol{A}$ has a repeated eigenvalue. In the case that the matrix $\boldsymbol{A}$ is real and symmetric, the system is decoupled, and the solution is trivial. For instance, take for $\boldsymbol{A} = \begin{pmatrix} -1 & 0 \\ 0 & -1 \end{pmatrix}$, that has a repeated eigenvalue $\lambda = -1$. Two eigenvectors are $\boldsymbol{\xi}_1 = (1, 0)$ and $\boldsymbol{\xi}_2 = (0, 1)$. As these are linearly independent, the general solution is given by

$$\boldsymbol{x} = \begin{pmatrix} c_1 \\ c_2 \end{pmatrix} e^{-t} \tag{4.24}$$

However, if the coefficient matrix is not Hermitian, there may be fewer than 2 independent eigenvectors corresponding to an eigenvalue. In that case, the general solution is given by

$$c_1\boldsymbol{\xi}_1 e^{\lambda t} + c_2(\boldsymbol{\xi}_1 t e^{\lambda t} + \boldsymbol{\xi}_2 e^{\lambda t}) \tag{4.25}$$

where the second vector $\boldsymbol{\xi}_2$ is any solution of

$$(\boldsymbol{A} - \lambda\boldsymbol{I})\boldsymbol{\xi}_2 = \boldsymbol{\xi}_1. \tag{4.26}$$

For more details, see e.g. [16].

4.2.4 *Classification of Fixed Points*

The previous examples show that for systems of linear equations, where the origin is always a fixed point, the eigenvalues $\lambda_{1,2}$ define the characteristics of the fixed points. From (4.8), we obtained the eigenvalues

Table 4.1 Characteristics of fixed points. A spiral is also denoted as a focus; a spiral sink is thus the same as a stable focus; a spiral source is also called an unstable focus

$\Delta < 0$		$\Delta > 0$ and $\tau^2 - 4\Delta > 0$	$\Delta > 0$ and $\tau^2 - 4\Delta < 0$	$\tau^2 - 4\Delta = 0$	$\Delta = 0$
saddle	$\tau < 0$	Stable node (sink)	Stable spiral (spiral sink)	Degenerate node (sink)	Lines of stable fixed points
	$\tau > 0$	Unstable node (source)	Unstable spiral (spiral source)	Degenerate node (source)	Lines of unstable fixed points
	$\tau = 0$	Center	Center		

$$\lambda_{1,2} = \frac{\tau \pm \sqrt{\tau^2 - 4\Delta}}{2}. \tag{4.27}$$

Check that we can also write $\Delta = \lambda_1\lambda_2$ and $\tau = \lambda_1 + \lambda_2$ for the determinant and trace, respectively. Depending on the determinant, $\Delta = \lambda_1\lambda_2$ and the trace $\tau = \lambda_1 + \lambda_2$, we can now classify the characteristics of the fixed points, as indicated in Table 4.1. Let us discuss this categorization in more detail. We will start with the situation where the determinant is smaller than zero, i.e. $\Delta < 0$. This implies that both eigenvalues are real, with one eigenvalue positive and the other negative, as was also the case in Example 4.4. One eigensolution will grow exponentially, the other will decay exponentially, and the fixed point is a saddle point. What happens when the $\Delta = 0$? In this case, we only have a single, real, eigenvalue (recall that $\Delta = \lambda_1\lambda_2$), and the general solution now evolves in a single direction in the x-y plane, either increasing (if the eigenvalue is positive) or decreasing, if the eigenvalue is negative. The solutions, therefore, are *lines of fixed points*, where the angle of the lines is defined by the two eigenvectors (see Exercise 4.6).

Let us now consider the situation where the determinant, $\Delta > 0$, and $\tau^2 - 4\Delta > 0$. In that case, the eigenvalues are both real and either both positive ($\tau > 0$) or both negative ($\tau < 0$). If $\tau > 0$, the fixed point is a source and if $\tau < 0$ it is a sink. If $\tau = 0$, the system will display a periodic motion, and a fixed point does not exist. If the determinant, $\Delta > 0$ and $\tau^2 - 4\Delta < 0$ the eigenvalues are complex conjugates, and the equilibrium is a spiral sink if $\tau < 0$ (in that case both real parts are negative) or a spiral source if the sum of the eigenvalues, $\tau > 0$: the real parts of the eigenvalues are positive. In Fig. 4.2 we show the fixed points for the various possibilities of the eigenvalues in a graphical manner. We finally summarize all the possibilities for the solutions to (4.2) as follows:

- If $\boldsymbol{A}$ has two real and distinct eigenvalues, λ_i, with corresponding eigenvectors $\boldsymbol{\xi}_i$, $i = 1, 2$, then every solution of (4.2) is of the form

$$c_1\boldsymbol{\xi}_1 e^{\lambda_1 t} + c_2\boldsymbol{\xi}_2 e^{\lambda_2 t}. \tag{4.28}$$

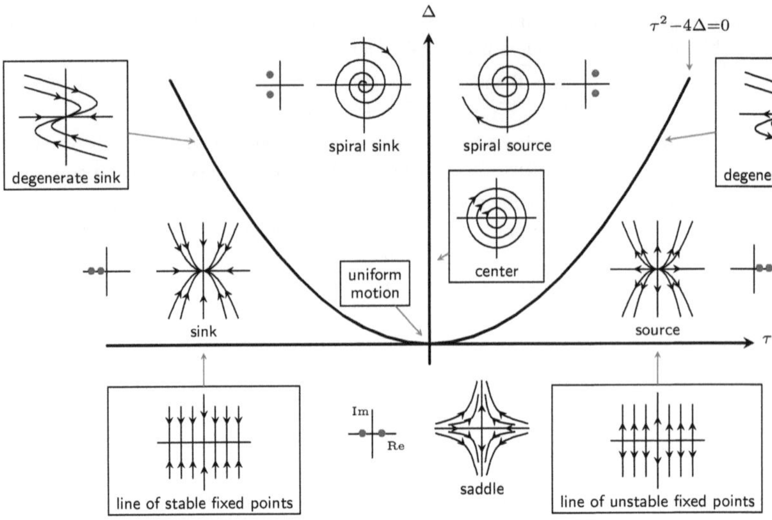

Fig. 4.2 The determinant—trace ($\Delta - \tau$) plane is divided into 5 regions according to the location of the eigenvalues (indicated with the red dots in the complex plane). If $\Delta < 0$, the eigenvalues are real and have opposite sign; hence the fixed point is a saddle point. For a spiral sink (stable spiral), both eigenvalues have a negative real part but opposite imaginary parts. Note that for all stable fixed points the eigenvalues are always in the left half-plane with Re $\lambda < 0$

- If A has a pair of complex conjugate eigenvalues, $\lambda = \rho \pm i\omega$, with eigenvectors $\boldsymbol{\xi}_1 = \boldsymbol{a} + i\boldsymbol{b}$ and $\boldsymbol{\xi}_2 = \boldsymbol{a} - i\boldsymbol{b}$, where $\boldsymbol{a}$ and $\boldsymbol{b}$ are real, the solution is given by

$$\boldsymbol{x} = c_1\boldsymbol{u}(t) + c_2\boldsymbol{v}(t) \tag{4.29}$$

with

$$\begin{aligned} \boldsymbol{u}(t) &= e^{\rho t}(\boldsymbol{a}\cos\mu t - \boldsymbol{b}\sin\mu t), \\ \boldsymbol{v}(t) &= e^{\rho t}(\boldsymbol{a}\sin\mu t - \boldsymbol{b}\cos\mu t). \end{aligned} \tag{4.30}$$

- if $\tau^2 - 4\Delta = 0$ then A has a repeated eigenvalue. If the matrix A is real and symmetric, the system was decoupled, and the solution is trivial. However, if we have only one linearly independent eigenvector (the matrix is defective), we must search for an additional solution. The general solution is given by

$$c_1\boldsymbol{\xi}_1 e^{\lambda t} + c_2(\boldsymbol{\xi}_1 t e^{\lambda t} + \boldsymbol{\xi}_2 e^{\lambda t}) \tag{4.31}$$

4.2.5 Drawing Solutions in the Plane

To obtain an impression of how our system behaves near an equilibrium point, it is often very helpful to sketch how the system will change near an equilibrium point starting from different initial conditions. We will illustrate this with a simple example.

Example 4.3 Let us solve the linear system given by

$$\dot{\boldsymbol{x}} = A\boldsymbol{x} \tag{4.32}$$

with $A = \begin{pmatrix} a & 0 \\ 0 & -2 \end{pmatrix}$. This is a very simple case, as the equations for x and y are already decoupled. Check that the eigenvalues $\lambda_1 = -2$ and $\lambda_2 = a$, resulting in the general solution

$$\begin{aligned} x(t) &= x_0 e^{at} \\ y(t) &= y_0 e^{-2t}. \end{aligned} \tag{4.33}$$

We can now draw the *phase portrait* of this system for different values of a near the fixed point. Each point in the x-y phase plane corresponds to a particular initial condition (x_0, y_0), and the change as a function of time is given by (4.33). If we assume that the parameter $a < 0$, both x and y will move towards the origin. If we take $a > 0$, starting at any value $x \neq 0$ will move x away from the origin, while any point $y \neq 0$ will still draw it closer to the origin. The origin, therefore, is a *saddle node*. This is graphically illustrated in Fig. 4.3.

* * *

In this example, where the two equations were decoupled, the eigenvectors were in the same direction as the x- and y-axis. This is not generally the case as the matrix that operates on $\boldsymbol{x}$ essentially performs scaling, stretching and rotation as possible operations.[2] Let us illustrate this with another example.

Example 4.4 We wish to solve

$$\dot{\boldsymbol{x}} = \begin{pmatrix} 1 & 2 \\ 4 & -1 \end{pmatrix} \boldsymbol{x}. \tag{4.34}$$

[2] Recall that a linear transform essentially scales an area, including compression, and can reverse the orientation of the region. For instance $\begin{pmatrix} -1 & 1 \\ 1 & 1 \end{pmatrix}$ rotates 45°.

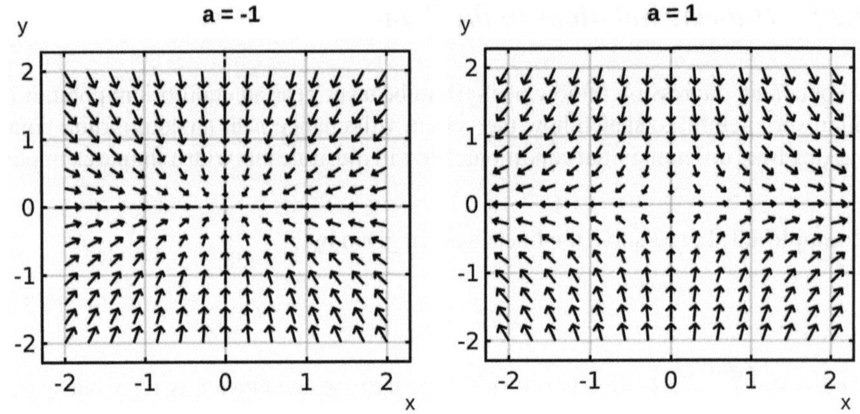

Fig. 4.3 Phase portraits showing vector fields of the system of (4.32). left: $a < 0$. The origin is a stable node. All trajectories move toward the origin. Right: $a > 0$, resulting in a saddle point

The characteristic equation is $\lambda^2 = 9$, that we solve to obtain the eigenvalues. We obtain $\lambda_1 = 3, \lambda_2 = -3$. The corresponding eigenvectors $\boldsymbol{v} = (v_1, v_2)$ need to satisfy

$$\begin{pmatrix} 1-\lambda & 2 \\ 4 & -1-\lambda \end{pmatrix} \begin{pmatrix} v_1 \\ v_2 \end{pmatrix} = \begin{pmatrix} 0 \\ 0 \end{pmatrix}. \tag{4.35}$$

For $\lambda_1 = 3$ the eigenvector $\boldsymbol{v_1} = (1, 1)$ or any scalar multiple thereof. For $\lambda_2 = -3$ we obtain $\boldsymbol{v_2} = (1, -2)$. The general solution of (4.34) is now given by

$$\boldsymbol{x} = c_1 \begin{pmatrix} 1 \\ 1 \end{pmatrix} e^{3t} + c_2 \begin{pmatrix} 1 \\ -2 \end{pmatrix} e^{-3t}. \tag{4.36}$$

Using the eigenvectors and eigenvalues we can now easily sketch the phase portrait. As $\lambda_1 = 3$ the first eigensolution grows exponentially, and as $\lambda_2 = -3$ the second eigensolution decays exponentially. Therefore, the origin is a *saddle point*. The two eigenvectors and the phase portrait are shown in Fig. 4.4.

* * *

4.3 Nonlinear Autonomous Differential Equations in the Plane

In biological systems, we typically need nonlinear equations as a representative model of the reality. For instance, the Hodgkin-Huxley equations that we discussed in Chap. 1 describe the generation of the action potential are nonlinear. While for

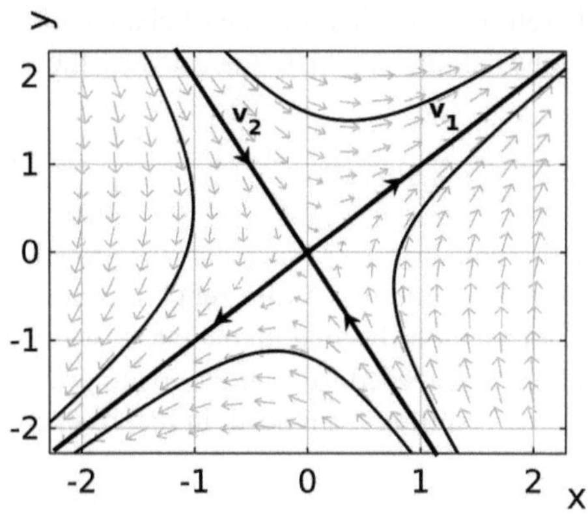

Fig. 4.4 Phase portrait the system of (4.34). The two eigenvectors are indicated with $\boldsymbol{v}_1$ and $\boldsymbol{v}_2$ with eigenvalues $\lambda_1 = 3$ and $\lambda_2 = -3$, respectively. The origin is a saddle point. As with all saddle points, a trajectory approaches the unstable manifold as $t \to \infty$ and the stable manifold as $t \to -\infty$. Note that we set $\boldsymbol{x} = (x, y)$

the two-dimensional linear equations, we can often find an analytical solution,[3] this is generally not possible for a system of nonlinear differential equations. However, as we will discuss in Sect. 4.4, we can often obtain a global understanding of the dynamics by using a graphical analysis. Before we turn to that technique, we will first discuss a classical example from biology that is described with an autonomous system of two nonlinear differential equations.

Example 4.5 The example we will discuss concerns two competing species. One of the animals (say rabbits) can find ample food at all times, but the other species, say foxes, need the rabbits as their sole source of food. Second, assume that the foxes' food depends entirely on the size of the rabbit population. Third, the rate of change of either the foxes or the rabbits is proportional to their size. Fourth, the environment is static, and the animals will not change their characteristics. Finally, the foxes have limitless appetite. We can now write for the change in the size of the population of rabbits

$$\frac{dx}{dt} = \alpha x - \beta xy. \tag{4.37}$$

The equations that describe this "predator prey problem" were independently proposed and studied by Lutka and Volterra in the early nineties of the previous century. If there were no foxes, the rabbit population with an unlimited food supply will reproduce exponentially, represented in the equation above by the term αx, unless subject to predation. We further assumed that the rate of predation upon the prey is proportional to the rate at which the predators and the prey meet, scaled with β. As meeting the foxes will reduce the size of the rabbit population, we subtract the term βxy. If either population is zero, then there can be no predation. The equation

---

[3] Some argue that we can always find an analytical solution for ordinary linear differential equations. However, if the system has variable coefficients this can be debated as the solution may contain integrals. For instance, $t^2y' + \sin(t)^2 y = 0$ has solution $C \exp\left(1/t - \cos(t)^2/t - \int_0^x \sin(t)/t\right)$.

therefore states that the rate of change of the rabbits is given by its own growth rate minus the rate at which it is eaten by the foxes. The equation for the foxes becomes

$$\frac{dy}{dt} = \delta xy - \gamma y. \tag{4.38}$$

In this equation, δxy represents the growth of the foxes, that is fully dependent on the presence of rabbits. A different constant is used, as the rate at which the foxes grows is not necessarily equal to the rate at which it consumes the rabbits (typically several rabbits are needed for a single fox to grow). The loss rate of the foxes is represented by γy, resulting from death in the absence of rabbits. So our system becomes

$$\begin{aligned} \dot{x} &= \alpha x - \beta xy \\ \dot{y} &= -\gamma y + \delta xy. \end{aligned} \tag{4.39}$$

We discussed earlier that the size of the populations may show periodic oscillations. In Exercise 4.10 you will study this in more detail.

4.3.1 Stability Analysis for Nonlinear Systems

Similar to the stability analysis for a one-dimensional system, we can apply this approach to two-dimensional nonlinear systems. Further, it provides information about how fast the approach towards a stable equilibrium is or the decay from an unstable equilibrium. While for the one-dimensional case we essentially stated the result, we will now more formally derive it. Our system of equations was given by

$$\begin{aligned} \frac{dx}{dt} &= f(x, y) \\ \frac{dy}{dt} &= g(x, y). \end{aligned} \tag{4.40}$$

If (4.40) has an equilibrium at the point (x_0, y_0) we can put it at the origin by introducing

$$\begin{aligned} \xi &= x - x_0 \\ \eta &= y - y_0. \end{aligned} \tag{4.41}$$

We thus obtain the system of equations

$$\begin{aligned} \frac{d\xi}{dt} &= f(\xi + x_0, \eta + y_0) \\ \frac{d\eta}{dt} &= g(\xi + x_0, \eta + y_0). \end{aligned} \tag{4.42}$$

If we make the Taylor series expansion of the right hand side and only retain the linear terms in ξ and η, then we obtain the so-called *linearisation* of (4.41) about this equilibrium.

$$\begin{aligned}\frac{d\xi}{dt} &= a\xi + b\eta \\ \frac{d\eta}{dt} &= c\xi + d\eta,\end{aligned} \tag{4.43}$$

where

$$\begin{aligned} a &= \frac{\partial f}{\partial x}(x_0, y_0); \quad b = \frac{\partial f}{\partial y}(x_0, y_0) \\ c &= \frac{\partial g}{\partial x}(x_0, y_0); \quad d = \frac{\partial g}{\partial y}(x_0, y_0). \end{aligned} \tag{4.44}$$

Hence, the disturbance evolves according to

$$\frac{d}{dt}\begin{pmatrix}\xi \\ \eta\end{pmatrix} = A\begin{pmatrix}\xi \\ \eta\end{pmatrix} \tag{4.45}$$

with

$$A = \begin{pmatrix} \frac{\partial f}{\partial x}(x_0, y_0) & \frac{\partial f}{\partial y}(x_0, y_0) \\ \frac{\partial g}{\partial x}(x_0, y_0) & \frac{\partial g}{\partial y}(x_0, y_0) \end{pmatrix}. \tag{4.46}$$

is called the Jacobian matrix at the fixed point (x_0, y_0), and is the multivariate analog of the one-dimensional case discussed in the previous chapter. Using the Jacobian now defines the nature of the equilibrium at the fixed points similar to the analysis for linear systems. We can now use the Jacobian to evaluate the stability of fixed points, using the approach discussed for the linear case.

Example 4.6 Let us illustrate how to determine he stability of the fixed points for the predator prey problem, (4.39), repeated here for convenience

$$\begin{aligned}\dot{x} &= \alpha x - \beta x y \\ \dot{y} &= -\gamma y + \delta x y.\end{aligned} \tag{4.47}$$

For the Jacobian matrix $\boldsymbol{J}(x, y)$ we obtain

$$J(x, y) = \begin{bmatrix} \alpha - \beta y & -\beta x \\ \delta y & \delta x - \gamma \end{bmatrix}. \tag{4.48}$$

When we evaluate this at fixed point $(0, 0)$ we obtain

$$J(0, 0) = \begin{bmatrix} \alpha & 0 \\ 0 & -\gamma \end{bmatrix} \tag{4.49}$$

with eigenvalues $\lambda_1 = \alpha, \quad \lambda_2 = -\gamma$. Since both $\alpha > 0$ and $\gamma > 0$ (by construction) the fixed point at the origin is a saddle point. This also implies that in this model the extinction of both species is not possible. Indeed, the populations of prey and predator can get infinitesimally close to zero and still recover. Evaluating the Jacobian at the second fixed point leads to

$$J\left(\frac{\gamma}{\delta}, \frac{\alpha}{\beta}\right) = \begin{bmatrix} 0 & -\frac{\beta\gamma}{\delta} \\ \frac{\alpha\delta}{\beta} & 0 \end{bmatrix}. \tag{4.50}$$

Note that the trace of this matrix, $\tau = 0$. The eigenvalues of this matrix are

$$\lambda_1 = i\sqrt{\alpha\gamma}, \quad \lambda_2 = -i\sqrt{\alpha\gamma}. \tag{4.51}$$

As the eigenvalues are both purely imaginary and conjugate to each others, this fixed point is elliptic, and in this particular case it it is a center.[4] Compare this with Table 4.1 and Fig. 4.2. This implies that the solutions are periodic and oscillating on a small ellipse around the fixed point.[5] As will be illustrated in Exercise 4.10 the curves are closed orbits surrounding the fixed point: the levels of the predator and prey populations oscillate without damping (the real parts of the eigenvalues are both zero).

4.4 Phase Plane Analysis

While linear differential equations can often be solved analytically,[6] this does not hold true for nonlinear equations. Starting with the general form of a two-dimensional system of autonomous differential equation

$$\begin{aligned} \frac{dx}{dt} &= f(x, y) \\ \frac{dy}{dt} &= g(x, y), \end{aligned} \tag{4.52}$$

we can show the global characteristics of the solutions of a system of differential equations with two independent variables as curves in the x-y plane. This may include the position and characteristics of fixed points, the presence of orbits (periodic solutions) and the direction of trajectories near fixed points or periodic orbits. In fact, we already did draw phase portraits near the fixed points for linear systems. While these

---

[4]In general, purely imaginary eigenvalues in a nonlinear system can show several behaviors.

[5]It is also possible to calculate the period of oscillation, that is given by $\omega = \sqrt{\lambda_1 \lambda_2} = \sqrt{\alpha\gamma}\omega = \sqrt{\lambda_1 \lambda_2} = \sqrt{\alpha\gamma}$.

[6]Linear differential equations with variable coefficients cannot always be solved directly, see e.g. [95].

phase portraits were generally simple, for nonlinear systems the phase portraits can be very complex, including a multiple of fixed points with different characteristics and (multiple) limit cycles (to be discussed in more detail in Sect. 4.5). An example of a phase portrait of a nonlinear system is shown in Fig. 4.5. Given our system of nonlinear equations, we wish to obtain a qualitative understanding of its behaviour. We proceed as follows. First, we wish to determine the points where fixed points exist, similar to our analysis of the linear one-dimensional systems discussed in the previous chapter. Recall that these points define the *equilibria* as solutions of the system of equations

$$\begin{cases} f(x,y) = 0 \\ g(x,y) = 0. \end{cases} \tag{4.53}$$

While in the one-dimensional situation we only had to satisfy one constraint ($\dot{x} = 0$), we now have *two* constraints, i.e. $\dot{x}$=0 and $\dot{y} = 0$. By solving the two equations, we can define the *x-nullcline* as the set of points where $f = 0$ and the *y-nullcline* as the set of points where $g = 0$. At the nullclines, the direction of the vector field is now defined: it is vertical for the x-nullcline, as it holds that $\frac{dx}{dt} = f(x,y) = 0$, implying that there is no change in the x-direction. Similarly, the vector field is horizontal at the y-nullcline. As the equilibria are defined as those points where both $\frac{dx}{dt} = 0$ and $\frac{dy}{dt} = 0$, *the fixed points are now defined by the intersection of the two nullclines.* The nullclines further divide the plane in connected regions or quadrants. In each quadrant the arrows of the vector field point into the same direction. Crossing a nullcline implies a change of quadrant: between left and right, if it is a x-nullcline; between top and bottom, if it is a y-nullcline. As a second step, we consider the right hand side (RHS) of (4.52), also called the *vector field*: at every point in the (x, y) plane, the vector $[f(x,y), g(x,y)]^T$ gives the direction and the magnitude of the rate of change of the vector $(x, y)^T$. A plot of these vectors (arrows) on a grid is called a *direction field*. It is a way to graphically show the behavior of all solutions for all initial conditions to a given system of differential equations without having to solve the system explicitly.[7] If you now start at a particular position (the initial condition) in a particular quadrant defined by the nullclines it is often possible to sketch a *trajectory* (or *flow*) in the vector field that shows the evolution of the system, represented by the relation between the two variables, x and y aided by the direction field. This is similar to our treatise of scalar differential equations in the previous chapter. Recall that we neglect the explicit dependence on time, as this is often less relevant to obtain a global understanding of the dynamics. Time dependence does become important if we wish to study processes like e.g. neuronal synchronization (see Chap. 5).

[7]This technique goes back to Euler (1707–1783) and is known as the *Euler method*.

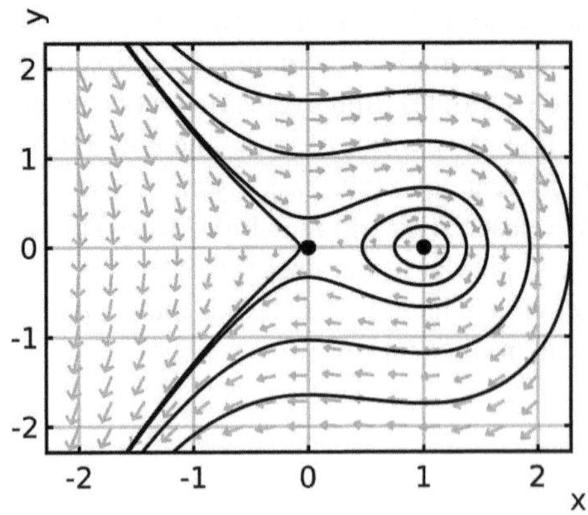

Fig. 4.5 Phase portrait of the system $\dot{x} = y$; $\dot{y} = x - x^2$ showing a few orbits, including periodic orbits (discussed in the Sect. 4.5). The solid circles indicate the two equilibria, (0, 0) and (1, 0)

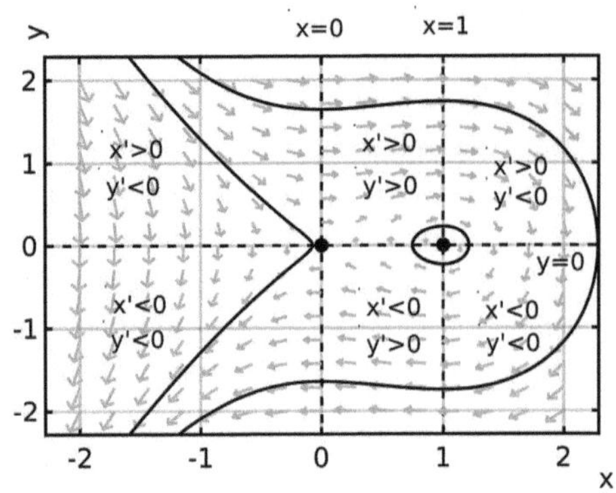

Fig. 4.6 Nullclines and vector field for the system of (4.54). The three nullclines divide the phase plane in 6 regions. For each region, we can determine the change in x and y. Given these directions, the trajectories can be sketched. The dotted line labeled y = 0 is the x-nullcline, $\dot{x} = 0$; the dotted lines labeled $x = 0$ and $x = 1$ are the two nullclines belonging to $\dot{y} = 0$

Example 4.7 For the system illustrated in Fig. 4.5

$$\begin{aligned} \dot{x} &= y \\ \dot{y} &= x - x^2 \end{aligned} \tag{4.54}$$

we wish to sketch solutions in the phase plane. We start by drawing the nullclines. The x-nullcline is given by $y = 0$ and the y-nullclines are given by $x = 0$ and $x = 1$, illustrated in Fig. 4.6. We can also derive the stability of the fixed points: the left equilibrium point is a saddle point; the right equilibrium is a spiral (cf. Table 4.1). By evaluating the RHS of the equations, we can subsequently draw the vector field and sketch possible solutions, as was already shown in Fig. 4.5.

4.5 Periodic Orbits and Limit Cycles

While we discussed equilibria, locations in the x–y plane where the system is at rest, we encountered another characteristic of 2D systems in Examples 4.5 and 4.7, i.e. they can show *periodic oscillations*, typically in a region surrounding a fixed point. Although both linear and nonlinear systems may show periodic oscillations, the characteristics in the phase plane are generally very different. If we observe a periodic oscillation in the phase plane, it holds for linear systems that neighboring points in the phase plane will also show periodic oscillations, and the amplitude of the oscillation is defined by the initial condition. If the amplitude is perturbed, the change of the amplitude will persist. Periodic oscillations in nonlinear systems, however, can exhibit *limit cycle* behavior: in the neighborhood of isolated closed trajectories in the phase plane, points will either spiral towards (stable or attracting limit cycle) or away (unstable limit cycle) from the limit cycle. A stable limit cycle, therefore, can *recover* from perturbations: if the system is moved away from the limit cycle, it will spiral towards and come infinitesimally close to its original orbit. This is very different from an unstable limit cycle where nearby points will spiral away. A graphic illustration of a stable and unstable limit cycle is shown in Fig. 4.7. Stable limit cycles are very common in nature. Breathing, for instance, could be conceptualized as limit cycle behavior. Typically, you are not aware of your breathing, and the system periodically allows air to enter and leave the airways with a frequency of approximately 12–14 breaths per minute. You can, however, deliberately increase or decrease its frequency. But if you let the system return to its default behavior, it will return to its baseline respiratory frequency, similar to what would be expected from stable limit cycle behavior. Other examples of stable limit cycle behavior include the beating of the heart, hormone secretion, and sleep-wake cycles. In pathology a limit cycle may become unstable or its region of attraction may change. These phenomena will be discussed in the Sect. 4.6, where we treat bifurcations. How can we establish if, given a particular nonlinear system, a limit cycle is present? The general answer is: we cannot. In some situations, it is possible to exclude that periodic solutions are present. Also, in some situations it can be proven that a closed orbit is present

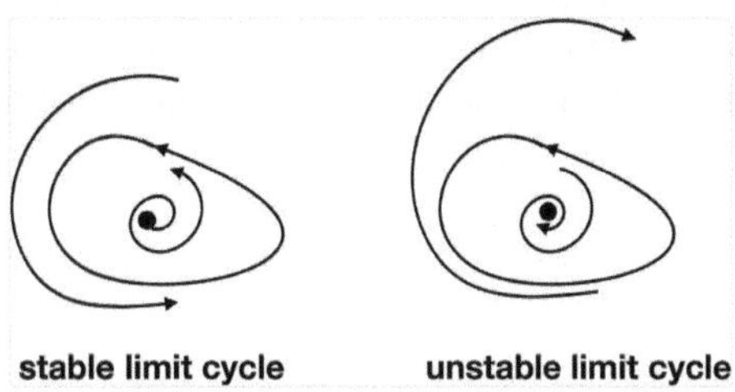

Fig. 4.7 Cartoon of a stable and unstable limit cycle. Left: In the neighborhood of the stable limit cycle, trajectories converge to the stable limit cycle (and the limit cycle surrounds an unstable fixed point). Right: In the neighborhood of an unstable limit cycle, trajectories move away from the limit cycle, and the limit cycle surrounds a stable fixed point (solid circle)

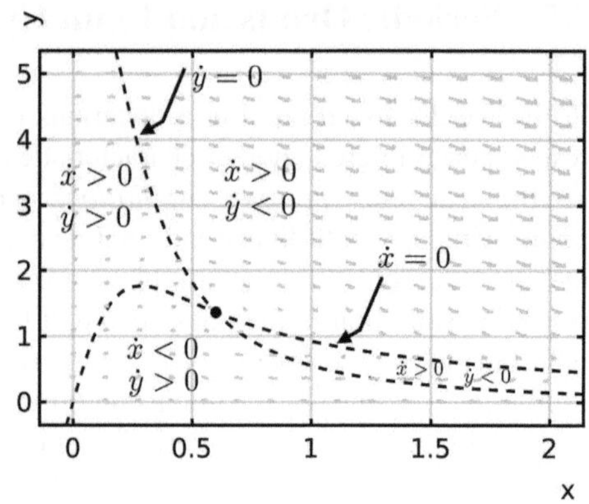

Fig. 4.8 Nullclines and direction field vectors of the Sel'kov model with a = 0.08, b = 0.6. As the fixed point is a repellor (spiral source) and the flow in a hypothetical space around the area of interest is directed inwards, a limit cycle exists

(Poincaré Bendixson theorem). We will not discuss this further, but we will show that by careful analysis of the phase portrait periodic solutions can be 'discovered'.

Example 4.8 Many processes in biology show oscillations. For instance, in the process of glycolysis, where ATP is generated. A model for these oscillations was proposed by Sel'kov:

$$\begin{aligned} \dot{x} &= -x + ay + x^2 y \\ \dot{y} &= b - ay - x^2 y \end{aligned} \tag{4.55}$$

with x and y the concentrations of ADP and F6P, respectively, and $a, b > 0$ kinetic parameters. You can get a good impression of the vector field by first plotting the nullclines and the fixed point in the phase plane. For $a = 0.08$ and $b = 0.6$ check that a spiral source exists at $x, y = (0.6, 1.36)$. Recall that at the nullcline $\dot{y} = 0$ the vector field is horizontal, and at the nullcline where $\dot{x} = 0$ it is vertical. By evaluating where the derivatives are positive or negative, you can now draw the vectors at the nullclines, as illustrated in Fig. 4.8. From that, you can now sketch the approximate flow of the limit cycle. You know it exists, as the fixed point is a repellor and a closed region around the repellor is present where the flow is *inward*: if you draw a hypothetical space around the phase plane shown, all arrows will *enter* this particular space. The actual limit cycle is shown in Fig. 4.9. Using pplane, you should verify that nearby initial conditions do indeed converge to the stable limit cycle.

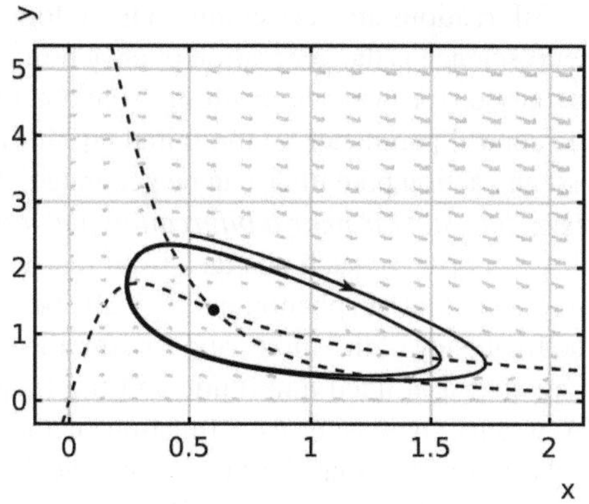

Fig. 4.9 Stable limit cycle in the Selkov model with a = 0.08, b = 0.6. Verify yourself that nearby initial points do indeed converge to the stable limit cycle shown

4.6 Bifurcations

Thus far we have discussed equilibria and how to evaluate their characteristics. In the case of linear equations, where the fixed point is always located at the origin of the phase plane, we use the characteristic equation from the matrix $\boldsymbol{A}$ given in (4.2). For a system of nonlinear equations, we discussed that we can estimate the stability from the linearization in a fixed point, using the Jacobian matrix.

Similar to the bifurcation analysis we presented in the former chapter, where we studied changes in the stability, appearance or disappearance of fixed points *in response to a change in a particular parameter*, we now proceed to how the dynamics of 2D systems may change. Recall the general expression for a system of (non-) linear equations, (4.52), but now let f and g depend on a parameter α, too. We therefore write:

$$\begin{aligned}\frac{dx}{dt} &= f(x, y, \alpha) \\ \frac{dy}{dt} &= g(x, y, \alpha).\end{aligned} \tag{4.56}$$

Instead of equilibria that were defined by particular values of (x, y) we may now observe *the loss or generation of new equilibria* depending on the parameter α, $x(\alpha)$ and $y(\alpha)$). For two dimensional systems, we will observe that other types of behavior than present in 1D may emerge, including *appearance or disappearance of oscillations*. A little more formal than in the previous chapter, we define that a bifurcation occurs if the phase portrait changes its topological structure, i.e. it loses its topological equivalence, as a parameter is varied. Recall that we can intuitively define topologically equivalence of two phase portraits if one is a distorted version of the other. For instance, for a 2D phase portrait drawn on paper, bending and warping of the paper preserves the topological equivalence, but not ripping: closed orbits must remain closed or trajectories connecting saddle points must not be broken.

Bifurcations are very common in biological systems, and the most familiar one in neuroscience is the generation of an action potential: if the input current to neuron is increased (for instance resulting from a net excitatory synaptic input, which acts as the control parameter α) the initial equilibrium (the resting membrane potential) is lost and action potentials can be generated. *A neuron is excitable, therefore, because its resting state is near a bifurcation, i.e. a transition to spiking.*

If we observe such changes from real-world measurements, different transitions may occur that can phenomenologically be differentiated from how the amplitude and the frequency of the spikes appear after the transition. For instance, spike amplitudes may start at a fixed amplitude and the frequency at a value significantly different from zero, increasing as a function of the current. In an alternative scenario, the spike frequency may be nearly zero at the bifurcation, gradually increasing as a function of the input current. If a change in a single parameter, here α, results in a bifurcation, the transition is a bifurcation of *codimension-1*. If systems depend on more, say m, parameters, bifurcations may occur if more than a single parameter is changed, and the transition is then of co-dimension-m. Here, we only discuss bifurcations of codimension-1.

Recall from Table 4.1 and Fig. 4.2 that if a two-dimensional system has a stable fixed point, both eigenvalues have a negative real part and are therefore located in the left-half of the complex plane. As the equilibria and associated eigenvalues for the system given by (4.56) now depend on the parameter α, it is possible that by varying the value of this parameter, the eigenvalues change, and therefore the equilibria. Let us assume that the system is at a stable fixed point. If both eigenvalues are real and negative, the fixed point is a stable node (sink). The other possibility is that the eigenvalues are complex conjugates and the fixed point is a stable spiral (spiral sink). If we now change the parameter α, two scenarios are possible. First, one of the real and negative eigenvalues passes the imaginary axis, i.e. $\lambda_1 = 0$ or $\lambda_2 = 0$. This results in a saddle node, transcritical or pitchfork bifurcation. Another possibility is that a pair of complex conjugate eigenvalues crosses the imaginary axis: this results in a Hopf bifurcation. A Hopf bifurcation occurs therefore if $\lambda_{1,2} = \pm i\omega$.

But how do we define the determinant and trace of our nonlinear system that now also depend on the parameter α? The answer is that we apply the same linearization as we applied before to determine the stability of a fixed point, but we now include the parameter α in our Jacobian:

$$\boldsymbol{A}(\alpha) = \begin{pmatrix} \frac{\partial f}{\partial x} & \frac{\partial f}{\partial y} \\ \frac{\partial g}{\partial x} & \frac{\partial g}{\partial y} \end{pmatrix} \Big\rfloor_{(x(\alpha), y(\alpha), \alpha)}. \tag{4.57}$$

If for a particular value of the parameter $\alpha = \alpha_0$ one of two real eigenvalues crosses the imaginary axis a saddle node, transcritical or pitchfork bifurcation occurs. If for a particular value of the parameter $\alpha = \alpha_0$ a pair of complex conjugate eigenvalues crosses the imaginary axis a Hopf bifurcation occurs. We will present examples of both types of bifurcations.

4.6.1 Saddle Node Bifurcation

We discussed this in the previous chapter for a 1D system. Here, we show an example in a planar system. Recall that in a saddle node bifurcation fixed points are either created or destructed. Let us consider the system

$$\begin{aligned} \dot{x} &= a - x^2 \\ \dot{y} &= -y. \end{aligned} \tag{4.58}$$

We first calculate the equilibria. If $a < 0$, no equilibria exist, for $a \geq 0$ we find that $x = \pm\sqrt{a}$ and $y = 0$. The eigenvalues for the equilibrium $(\sqrt{a}, 0)$ are $\lambda_1 = -1$ and $\lambda_2 = -2\sqrt{a}$. This equilibrium is a stable node (or sink) as both eigenvalues are real and negative (cf. Fig. 4.2). For the other equilibrium $(-\sqrt{a}, 0)$ the eigenvalues are $\lambda_1 = -1$ and $\lambda_2 = 2\sqrt{a}$, which implies that this equilibrium is a saddle. If we start with a value $a > 0$ and now reduce its value to $a = 0$ the two equilibria "merge"; the origin is now a double fixed point with eigenvalues $\lambda_1 = -1$ and $\lambda_2 = 0$. For $a < 0$, the equilibrium disappears. The phase plots for these three values of a are shown in Fig. 4.10.

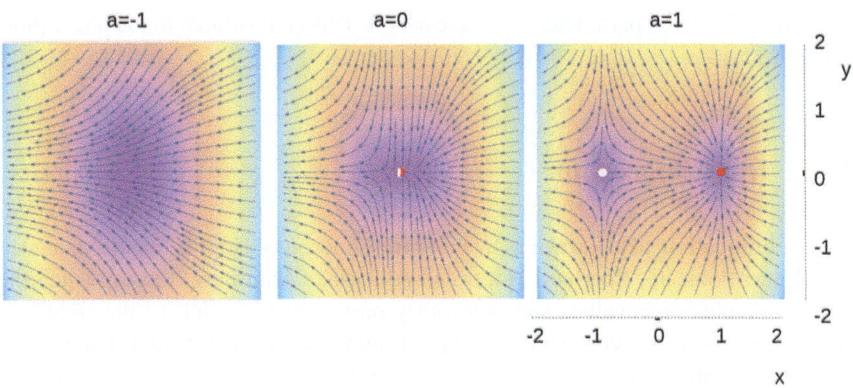

Fig. 4.10 Phase-plots of system given by (4.58) for three different values of the bifurcation parameter a. Slower field velocities are colour-coded with the purple colour. The stable node is indicated with a red dot; the saddle point with a white dot. For values $a < 0$ no equilibria exist. At $a = 0$ there is a saddle-node point. Here, the bifurcation occurs and two equilibria are created. Phase-plots created in Mathematica

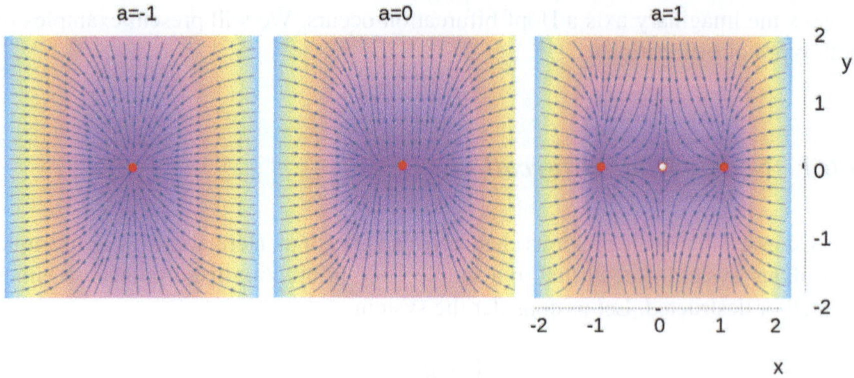

Fig. 4.11 Phase-plots of system given by (4.59) for three different values of the bifurcation parameter a. Slower field velocities are colour-coded with the purple colour. The stable fixed point (stable node) is indicated with a red dot; the unstable fixed point with a white dot. Phase-plots created in Mathematica

4.6.2 Supercritical Pitchfork Bifurcation

The supercritical pitchfork bifurcation occurs for the system

$$\begin{aligned} \dot{x} &= a - x^3 \\ \dot{y} &= -y \end{aligned} \tag{4.59}$$

when the bifurcation parameter has value $a = 0$. This is illustrated in the phase plots in Fig. 4.11.

4.6.3 Hopf Bifurcation

A Hopf bifurcation, also known as a (Poincaré-)Andronov-Hopf bifurcation, named after Henri Poincaré, Eberhard Hopf, and Aleksandr Andronov,[8] occurs when the pair of complex eigenvalues simultaneously moves from the left to the right half plane (cf. Fig. 4.2), the corresponding equilibrium loses it stability, and changes into an unstable spiral.[9] Let us for now assume that the flow in the phase space moves away from the equilibrium. What will happen next? There are two possible scenarios during this transition. An attracting periodic orbit (i.e. a stable limit cycle) *emerges from the equilibrium* as the parameter is increased beyond the bifurcation point, and

[8]Historically, Poincaré contributed to this topic (1892), Andronov and Witt discussed it around 1930 and Hopf's paper appeared in 1942. In the literature, it is also referred to as a 'Hopf bifurcation' or 'Andronov-Hopf' bifurcation.

[9]Of course, a movement from the right half to the left half plane is possible, too.

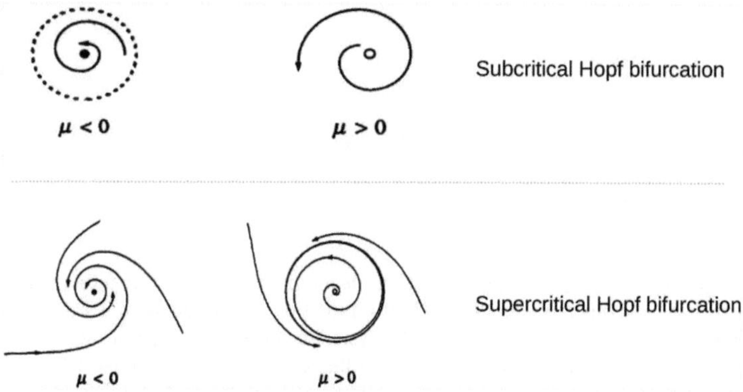

Fig. 4.12 Cartoon illustrating a subcritical Hopf (top) and a supercritical Hopf (bottom) bifurcation for a control parameter μ; the bifurcation occurs at $\mu = 0$. In the subcritical Hopf bifurcation an unstable limit cycle surrounds the equilibrium point. As the bifurcation is approached, the unstable limit cycle shrinks down to the equilibrium point, which becomes unstable in the process and large-amplitude oscillations occur. This contrasts with the supercritical Hopf bifurcation (bottom), where a stable limit cycle is born at the bifurcation and oscillations start at low amplitude, gradually increasing as the system moves further into the limit cycle regime

the system gradually evolves to a stable limit cycle. The stable limit cycle is initially very close to the original fixed point and grows slowly if the parameter moves further into the limit cycle regime. This type of Hopf bifurcation is known as a supercritical Hopf (or soft) bifurcation.

In the other scenario, we start with the situation that an unstable limit cycle encloses a stable fixed point. All initial conditions inside the unstable limit cycle will move towards the fixed point, and all initial conditions outside the unstable limit cycle will move away from the limit cycle, towards a distant attractor. The unstable limit cycle decreases in size when the bifurcation is approached, and at the bifurcation the limit cycle disappears. Now the orbit leaves the neighborhood of the equilibrium by jumping to a distant attractor, which may be a fixed point, another limit cycle or infinity: a subcritical or hard bifurcation. Both the subcritical and supercritical Hopf bifurcation are illustrated in Fig. 4.12. We can also sketch the bifurcation diagram that displays the amplitude of the oscillation for a sub- and supercritical Hopf bifurcation. In the case of a subcritical Hopf, the amplitude of the unstable oscillation gradually shrinks to zero as the system approaches the bifurcation. At the bifurcation the stable fixed point and the unstable oscillation collide and an unstable equilibrium emerges. For a supercritical Hopf, the stable fixed point becomes unstable, but at the same time a (small) stable limit cycle is born. This behaviour is sketched in Fig. 4.13.

In many situations, it is very important to differentiate between the two Hopf bifurcations. This is not limited to biology. Hopf bifurcations also occur in mechanical systems, aeronautics, and chemical reactions. In those fields, the name for the supercritical Hopf bifurcation is sometimes denoted as 'soft', 'continuous' or 'safe', while the subcritical Hopf bifurcation is referred to as 'hard', 'discontinuous' or

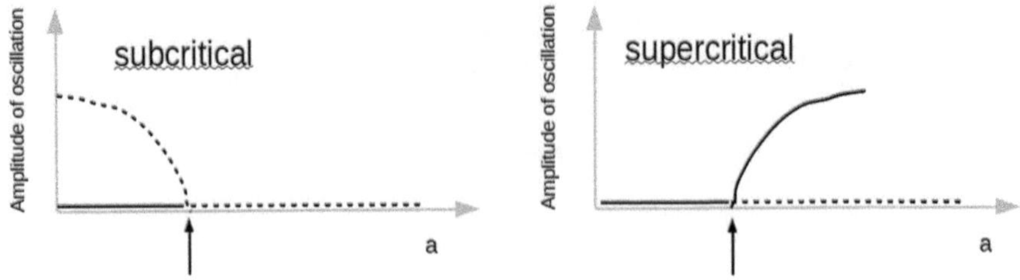

Fig. 4.13 Sketch of the amplitude of oscillations as a function of the bifurcation parameter a for a subcritical (left) and supercritical (right) Hopf bifurcation. The bifurcation is indicated with the vertical arrow. Note, that nonlinear terms of the system determine the type of bifurcation as these define the limit cycles and their stability. This is further exemplified in Exercise 4.14

'dangerous'. Indeed, as the bifurcation occurs in the subcritical Hopf bifurcation, it is not clear where the system will evolve to. This could be a distant limit cycle but even infinity. Conversely, when the supercritical Hopf bifurcation occurs, initially the system will evolve to a stable limit cycle with small amplitude.

While we can establish if a Hopf bifurcation occurs—a pair of complex eigenvalues crosses the imaginary axis-, how can we determine if the bifurcation is sub- or supercritical? Linear analysis via linearisation cannot distinguish the supercritical Hopf from the subcritical one: the bifurcation is always described by the nonlinear terms. This is visible in Fig. 4.13 as well: the linear part of the system for both conditions is the same: the stable fixed point with radius equal to zero changes from stable to unstable, but the limit cycle (that is described by a nonlinear term) defines the type of bifurcation. For simple systems, this can sometimes be calculated explicitly, as we will show by various examples later on and in some of the Exercises. Also, it is often possible to use numerical methods. If a small, attracting limit cycle appears immediately after the fixed point goes unstable, and if its amplitude shrinks back to zero as the parameter is reversed, the bifurcation is supercritical. If this is not the case, the bifurcation is probably subcritical and the nearest attractor might be far from the fixed point (cf. Fig. 4.13). More fundamentally, the particular type of bifurcation can be determined by using the *Lyapunov coefficient* that determines whether the bifurcation is soft (super-critical) or hard (sub-critical). This will not be discussed. For details, see e.g. [59].

Example 4.9 The two Hopf bifurcations can be demonstrated by the following system of equations:

$$\begin{aligned}\dot{x} &= \underbrace{\mu x + y}_{\text{linear}} + \underbrace{x(x^2+y^2)}_{\text{nonlinear}} \\ \dot{y} &= \underbrace{-x + \mu y}_{\text{linear}} + \underbrace{y(x^2+y^2)}_{\text{nonlinear}} \, .\end{aligned} \tag{4.60}$$

Note that these equations contain a linear part and a nonlinear part, as indicated. It is obvious that (0, 0) is a fixed point. Let us explore the stability. The Jacobian matrix is

$$\begin{pmatrix} \mu & 1 \\ -1 & \mu \end{pmatrix}. \tag{4.61}$$

The trace, $\tau = 2\mu$ and the determinant, $\Delta = \mu^2 + 1$. The eigenvalues $\lambda_{1,2} = \mu \pm i$. This implies that if $\mu < 0$ the fixed point is an asymptomatic stable spiral, and if $\mu >$ an unstable spiral. We rewrite this system in polar coordinates, $x = r\cos\theta$ and $y = r\sin\theta$. This implies that $r^2 = x^2 + y^2$. Differentiating both sides of this expression results in

$$r\dot{r} = x\dot{x} + y\dot{y}. \tag{4.62}$$

Inserting (4.60) in the RHS of (4.62), we obtain

$$\dot{r} = r(\mu + r^2). \tag{4.63}$$

If we evaluate the change in θ by taking the derivative of $\theta = \arctan\frac{y}{x}$, we obtain $\dot{\theta} = -1$. You can check this for yourself using $\dot{\theta} = \frac{1}{1+y^2/x^2}(\frac{x\dot{y}}{x^2} - \frac{y\dot{x}}{x^2})$ and inserting $\dot{x}$ and $\dot{y}$ using (4.60). We can now sketch r and θ in the x-y plane. For $\mu < 0$ an unstable limit cycle exists with $r = \sqrt{\mu}$, that surrounds the fixed point at the origin, which is an asymptotic stable spiral. For $\mu > 0$ there is only one non-stable spiral node at the origin. For $\mu = 0$, therefore, a subcritical Hopf bifurcation occurs. The solutions for the positive and negative values of μ are sketched in Fig. 4.14. If we change the sign of the nonlinear parts of the original equations, resulting in

$$\begin{aligned} \dot{x} &= \mu x + y - x(x^2 + y^2) \\ \dot{y} &= -x + \mu y - y(x^2 + y^2) \end{aligned} \tag{4.64}$$

we find that

$$\dot{r} = r(\mu - r^2). \tag{4.65}$$

For $\dot{\theta}$ it still holds that $\dot{\theta} = -1$. In this case, therefore, for $\mu > 0$ a stable limit cycle with radius $r = \sqrt{\mu}$ exists, with an unstable node (spiral) at the origin, while for $\mu < 0$ the origin is an asymptotically stable spiral. For $\mu = 0$, a supercritical Hopf bifurcation occurs. The solution is sketched in Fig. 4.15. As a final step, we can now draw the bifurcation diagram for the sub- and supercritical Hopf bifurcation. The bifurcation diagrams for both systems of the (4.60) and (4.64) are shown in Fig. 4.16.

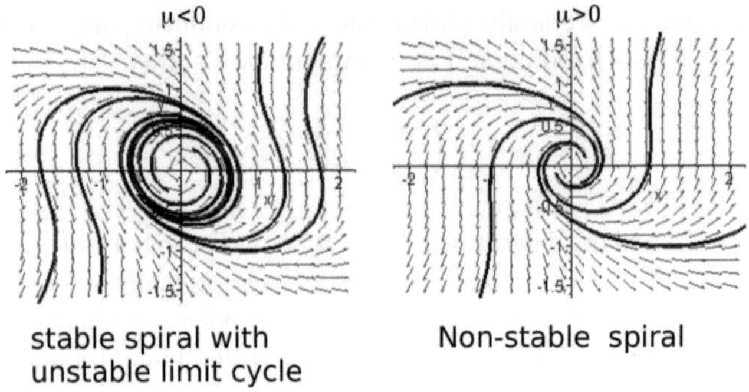

Fig. 4.14 Illustration of trajectories in the $x - y$ phase plane for the system given in (4.60) for negative and positive values for the control parameter μ. A subcritical Hopf bifurcation occurs at $\mu = 0$

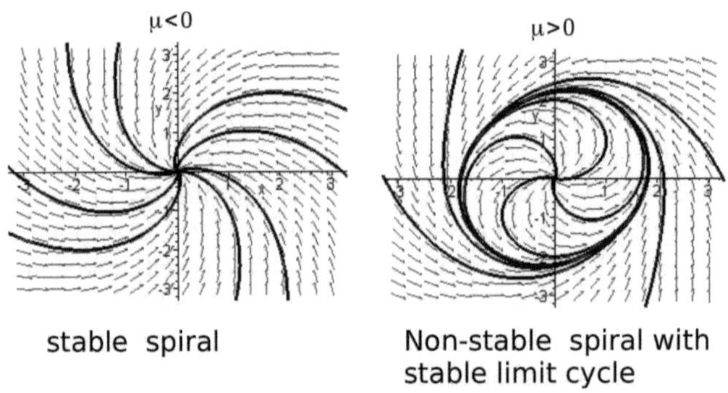

Fig. 4.15 Illustration of trajectories in the $x - y$ phase plane for the system given in (4.64) for negative and positive values for the control parameter μ. A supercritical Hopf bifurcation occurs at $\mu = 0$

4.6.4 Oscillations in Biology

In biological systems, a wealth of oscillations or *rhythms* can be observed on spatial scales ranging from individual neurons to neuronal populations. Further, oscillations generally change during pathology, including appearance and disappearance of pathological rhythms. For instance, if neurons are deprived of oxygen, their membrane potentials will change as the ATP dependent sodium potassium pump is not able to maintain the ionic gradients of sodium and potassium. At a critical point, spontaneous oscillations will occur as the neuron starts spiking: anoxic oscillations. This is an example of a bifurcation resulting from a change in the potassium (and sodium) gradients across the cell membrane. We will discuss these phenomena in more detail in Chap. 8.

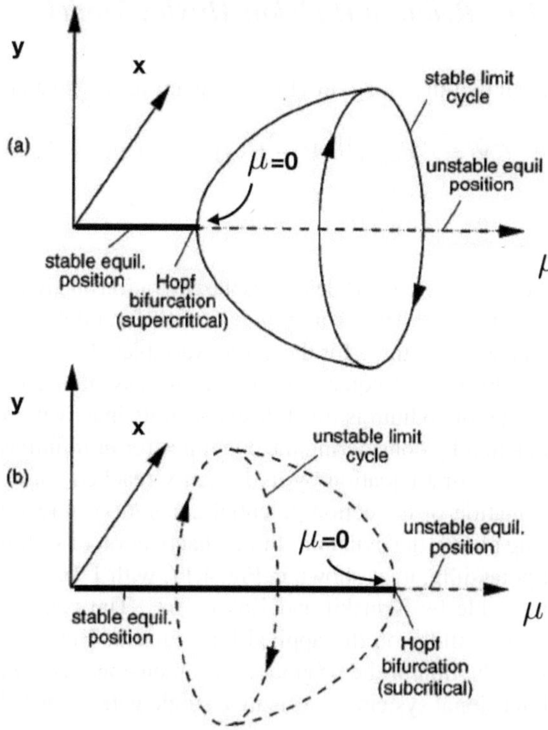

Fig. 4.16 Bifurcation diagram. Top: supercritical Hopf bifurcation. For values of $\mu < 0$ a stable spiral exists, that becomes unstable at $\mu = 0$, with the emergence of a stable limit cycle. Bottom: subcritical Hopf bifurcation: for values of $\mu < 0$ a stable spiral exists with an unstable limit cycle. The unstable limit cycle shrinks in size as the bifurcation is approached and the stable spiral disappears at $\mu = 0$ transits into an unstable spiral

4.7 Reductions to Two-Dimensional Models

We have shown that two-dimensional models can display a rich variety of dynamics. Many biological phenomena, however, are described by higher order models. For instance, the Hodgkin-Huxley equations form a four dimensional system of ordinary differential equations when viewed as a one compartmental model for the soma. In several situations, however, we can *reduce* a higher order system to a lower order system, with preservation of the key characteristics. Such a reduction is often both conceptually better to understand and also easier to visualize. We will illustrate this with a discussion of various three two-dimensional models for neurons with a voltage-gated sodium and a potassium channel. In these two-dimensional models we can relatively easy explore the effects of constant stimuli in the form of an applied bias current. As this gives us one parameter to vary, we may expect saddle-node bifurcations of equilibria and Hopf bifurcations to periodic solutions.

4.7.1 Reduced Hodgkin-Huxley Model

Recall that the Hodgkin-Huxley equations are given by

$$\begin{aligned} C\dot{V} &= -g_{Na}m^3h(V - E_{Na}) - g_K n^4(V - E_K) - g_L(V - E_L) + I \\ \dot{\chi} &= \frac{\chi_\infty(V) - \chi}{\tau_\chi(V)}, \end{aligned} \tag{4.66}$$

where $\chi \in \{m, n, h\}$ describes the activation and inactivation variables, as discussed in Sect. 1.3.4. This is a system of 4 differential equations: one for the membrane voltage $\dot{V}$ and three for the gating variables. It is possible to reduce this to a system of 2 differential equations, preserving essential characteristics. As the activation channel of sodium is much faster then its inactivation channel, we can apply a first reduction by considering the dynamics for m infinitely fast. We now simply replace m in the first equation by its instantly reached value $m_\infty(V)$. Second, if we run a simulation of the action potential generation in response to an external current, I, using (4.66), the dynamics of the inactivation of sodium h is similar to the activation of potassium, n, as shown in Fig. 4.17, with $1.1n + h \approx 0.89$. This observation was first made by Krinskii and Kokoz [67]. The constants that determine the line do depend mildly on the applied bias current. Here we take $h = 1 - n$. This results in a reduction of the original four dimensional Hodgkin-Huxley equations to a two dimensional system, now using a single gating variable n, only, expressed as:

$$\begin{aligned} C\dot{V} &= -g_{Na}m_\infty^3(V)(1 - n)(V - E_{Na}) - g_K n^4(V - E_K) - g_L(V - E_L) + I \\ \dot{n} &= \frac{n_\infty(V) - n}{\tau_n(V)} = \alpha_n(1 - n) - \beta_n n. \end{aligned} \tag{4.67}$$

The phase plot with the nullclines is shown in Fig. 4.18. In Exercise 4.17 we will explore some of its characteristics. Another model that reduces the Hodgkin-Huxley equations to two dimensions is the Morris-Lecar model, developed by Catherine Morris en Harold Lecar. With this model, it is possible to produce a variety of oscillatory patterns depending on conductances of the calcium and the potassium channel.

4.7.2 Morris-Lecar Model

The Morris–Lecar (ML) model also reduces the Hodgkin-Huxley equations from four dimensions to two, In this reduction, the model contains a calcium channel instead of the sodium channel. The model was developed by Catherine Morris and Harold Lecar to reproduce the variety of oscillatory behavior in relation to Ca^{++} and K^+ conductance in the muscle fiber of the giant barnacle. The Morris-Lecar model (and variations thereof) is also used as a prototype educational model to explain stability

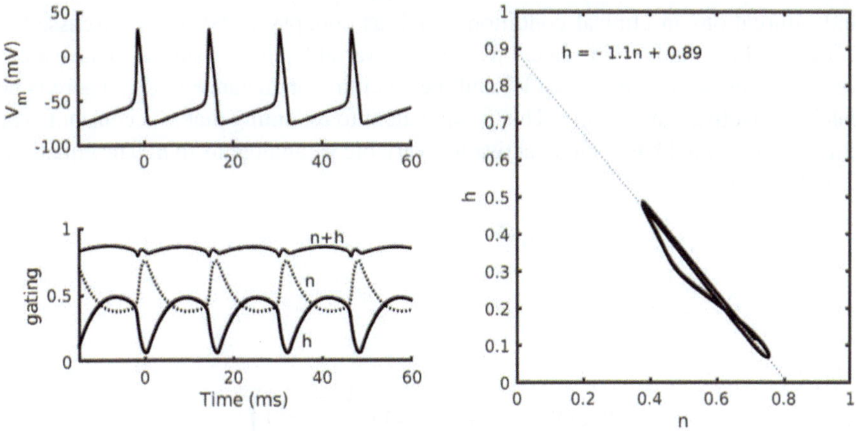

Fig. 4.17 Simulation of the action potential using the Hodgkin-Huxley equations with $I > 0$. Upper left panel: train of action potentials in response to the excitatory current. Left lower panel: dynamics of the h and n gating variables. Note that their sum is nearly 1 and constant: $1.1n + h \approx 0.89$. The panel on the right shows their linear dependency

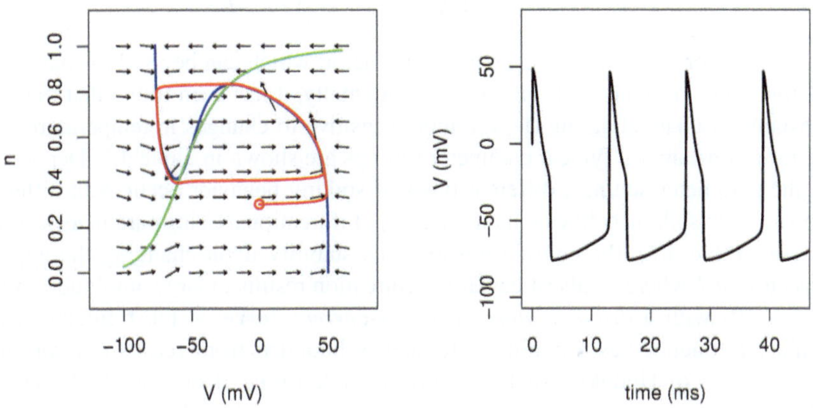

Fig. 4.18 Left: Phase-plane for the reduced Hodgkin-Huxley model (4.67). A limit cycle is shown in red. The V and w-nullclines are indicated in blue and green, respectively. The constants used are: $g_{Na} = 120$ mS/cm^2, $g_K = 36$ mS/cm^2, $E_{Na} = 50$ mV, $E_K = -77$ mV, $E_L = -54.4$ mV, $g_L = 0.3$ mS/cm^2, $I = 10$ pA, $\alpha_n = \frac{0.01(V+55)}{1-\exp(-(V+55)/10)}$, $\beta_n = 0.125\exp(-(V+65)/80)$, $\alpha_m = \frac{0.1(V+40)}{1-\exp(-(V+40)/10)}$, $\beta_m = 4\exp(-(V+65)/18)$ and C=1 μF/cm^2. Right: action potentials. Note the small distortion of the shape of the down-slope of the action potential, resulting from the approximation

and bifurcations in clinical conditions, such as epilepsy [108], to be discussed in more detail in Chap. 9. The calcium activation variable, m, is assumed to have a fast time constant so that $m = m_\infty(V)$ and the calcium conductance is g_{Ca}. There is no calcium inactivation variable. This is equivalent to assuming that h is constant. The potassium channel has a single activation variable w, analogous to n. The equations take the form

$$\begin{aligned} C\dot{V} &= -g_{\text{Ca}} m_\infty(V)(V - E_{\text{Ca}}) - g_K w(V - E_K) - g_L(V - E_L) + I \\ \dot{w} &= \phi \frac{w_\infty(V) - w}{\tau_w(V)}, \end{aligned} \tag{4.68}$$

where

$$\begin{aligned} m_\infty(V) &= \frac{1}{2}\left(1 + \tanh(\frac{V - V_1}{V_2})\right) \\ \tau_w(V) &= \frac{1}{\cosh(\frac{V-V_3}{2V_4})} \\ w_\infty(V) &= \frac{1}{2}\left(1 + \tanh(\frac{V - V_3}{V_4})\right). \end{aligned}$$

The parameter ϕ in (4.68) is a temperature factor, which can be used to change the relative time constants of V and w. Experimentally, it has been found that the time constants of channel gating ($\dot{w}$) are more sensitive to changes in temperature than the time constant $\dot{V}$. Typical parameter settings are shown in Table 4.2. Depending on the parameter settings, different types of spiking behavior occur with different transitions to spiking behavior, including Hopf bifurcations. Using parameter settings S1 (see Table 4.2), the equilibrium loses its stability upon changing the applied bias current I where a subcritical Hopf bifurcation results in large amplitude spikes (Fig. 4.19) with a *discontinuous frequency-current curve*: at the bifurcation, the initially frequency has a minimum frequency bounded from zero. In the original classification of Hodgkin, such a neuron is called type II or class II. Using the other parameter setting (S2), a periodic orbit appears when the equilibrium loses its stability, with a frequency that can be arbitrarily small. This is called class I or type I behaviour and the frequency-current curve is continuous. This is a saddle-node-onto-limit-cycle bifurcation. We will study the ML model and the bifurcations in more detail in Chap. 9 about epilepsy and in Problem 4.18. Another reduction of the Hodgkin-Huxley equations was proposed by Fitzhugh and independently by Nagumo.

4.7.3 Fitzhugh-Nagumo Model

The Fitzhugh-Nagumo (FHN) model is a simple model to generate action potentials, and has only two dynamic variables, making it possible to explore the dynamics using

Table 4.2 Two different parameter settings for the Morris Lecar Model, resulting in type I or type II behavior

Parameter	Setting 1 (S1)	Setting 2 (S2)	Unit
g_{Ca}	4.4	4.4	mS/cm^2
g_{K}	8.0	8.0	mS/cm^2
g_{l}	2.0	2.0	mS/cm^2
C	20	20	μF/cm^2
E_{Ca}	120	120	mV
E_{K}	−84	−84	mV
E_{L}	−60	−60	mV
ϕ	0.04	0.0667	–
V_1	−1.2	−1.2	mV
V_2	18	18	mV
V_3	2	12	mV
V_4	30	17	mV

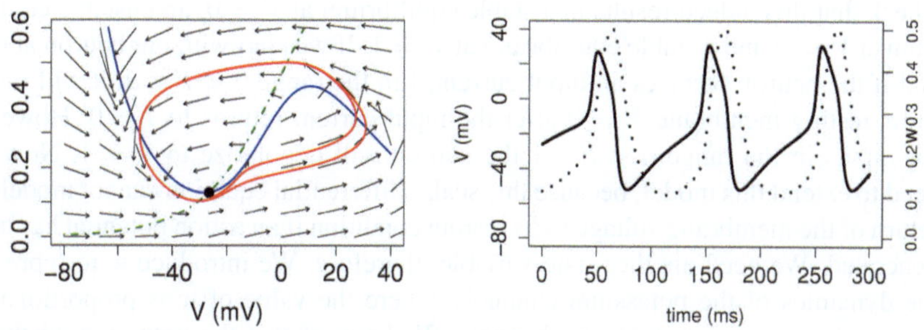

Fig. 4.19 Left: phase plane for the ML equation with an external current $I = 90$ μA/cm^2 and initial condition $V(0) = -26$ mV and $w(0) = 0.1134$. The parameters used are $\phi = 0.04$, $V_1 = -1.2$ mV, $V_2 = 18$ mV, $V_3 = 2$ mV, $V_4 = 30$ mV, $E_{Ca} = 120$ mV, $E_K = -84$ mV, $E_L = -60$ mV, $g_K = 8$ mS/cm^2, $g_{Ca} = 4.4$ mS/cm^2, $g_L = 2$ mS/cm^2, $C = 20$ μF/cm^2. The nullclines are shown in blue and green, respectively. The orbit, starting at $V(0) = -26$ mV, $w(0) = 0.1134$, converges to the periodic orbit. Right: time course of the membrane voltage V (solid line) and the variable w (dotted line).

phase plane methods. It is not so difficult to derive the FHN model starting from a very simple cell, containing sodium and potassium channels, only, each controlled by a voltage-gated channel.[10] We will show that if we simplify the gating of these channels, we can obtain a system of only two differential equations that simulates the generation of an action potential.

[10]This treatise is strongly motivated by "A simple spiking neuron mode: sodium and potassium channels. Nykamp DQ. From Math Insight". http://mathinsight.org/video/simple_spiking_neuron_model_sodium_potassium_channels.

Let's start with a simple model neuron, that only contains voltage-gated sodium channels. The resting membrane potential is at a particular value, V. We can increase the membrane voltage, which will open voltage gated sodium channels. The membrane potential will therefore become less negative, tending towards the Nernst potential of sodium, at the same time opening more sodium channels (as a result of the positive feedback loop). We further assume that a particular threshold voltage is needed in order for the sodium channels to open, and that for small changes in the membrane potential, the membrane will return to its resting value. We can combine these three properties in a simple autonomous differential equation with properties that (i) the voltage increases when sodium channels open; (ii) existence of a positive feedback loop and (iii) a threshold voltage. Let us further set the resting potential of the membrane to zero and the voltage of the action potential to one, and the threshold voltage, a in the range $[0, 1]$. The simplest relation between the membrane voltage and the opening of the voltage-gated sodium channels can therefore be described as a cubic function:

$$\dot{V} = -V(V - a)(V - 1). \tag{4.69}$$

Check that this indeed results in a stable equilibrium at $V = 0$, an unstable equilibrium at $V = a$ and a stable equilibrium at $V = 1$. If we start with our neuron at rest, and if the neuron receives an input current, I in the range $0 < I < a$, it will return to its resting membrane $V = 0$ after the input current returns to $I = 0$. However, for inputs in the range $a < I < 1$ the neuron will depolarize to $V = 1$. Now we need to extend this model, because this scalar differential equation cannot model the return of the membrane voltage to its resting condition if an action potential has been generated. We need another state variable, therefore. We introduce w to represent the dynamics of the potassium channels, where the value of w is proportional to the number of open potassium channels. We know that if the potassium channels open in response to an increase in membrane voltage, the membrane potentials tends to reduce to its resting value. Further, potassium channels are relatively slow. The dynamics of the potassium channels can for now be modeled by

$$\dot{w} = \varepsilon(V - \gamma w) \tag{4.70}$$

with $\gamma > 0$. Note, that for a fixed value of the voltage V that $w = V/\gamma$ is a stable fixed point; therefore, w will evolve towards its value $w = V/\gamma$. As potassium channel dynamics is a relatively slow process, we introduce ε as a small number, that multiplies the 'moving target' V/γ. We will now combine the sodium and potassium channel kinetics to allow return of the membrane potential to zero after an action potential has been generated. Our initial equation for the membrane potential will therefore also depend on the variable w, and in its simplest form we write for our dynamical system

$$\begin{aligned} \dot{V} &= V(1-V)(V-a) - w + I \\ \dot{w} &= \varepsilon(V - \gamma w), \end{aligned} \tag{4.71}$$

where we simply subtracted w from the expression for the membrane voltage as an increase in the value of w will *reduce* the membrane potential, as we just argued. Further, we add I as an external current. Recall, that the value of $a \in (0, 1)$ is the 'threshold voltage', $\gamma \geq 0$ defines how strong w depends on the membrane voltage and $0 < \varepsilon \ll 1$ defines how fast w responds to the membrane voltage. These equations are known as the Fitzhugh-Nagumo equations that have reduced the four Hodgkin Huxley equations to a planar system. The nullclines are given by $w = V(1-V)(V-a) - I$ (V-nullcline), $w = V/\gamma$ (w-nullcline), as shown in Fig. 4.20. In Exercise 4.19 we examine this system in some more detail. In the literature, you may find other formulations of the Fitzhugh-Nagumo models, for instance

$$\begin{aligned} \dot{V} &= V - \frac{V^3}{3} - w + I \\ \dot{w} &= \varepsilon(V - aw + b), \end{aligned} \tag{4.72}$$

where V represents the membrane voltage and w a recovery variable. I is the applied current and a and b are non-negative parameters and $0 < \varepsilon \ll 1$ defines the timescale of the slow recovery parameter w. The V-nullcline is given by $w = V - V^3/3 + I$ and the w-nullcline is given by $w = (b + V)/a$.

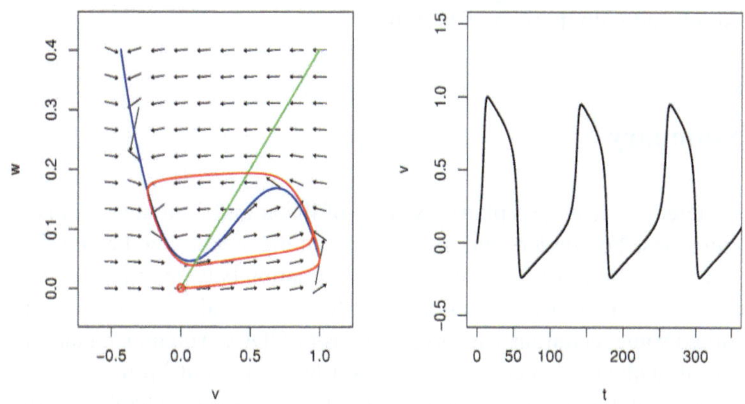

Fig. 4.20 The FitzHugh-Nagumo (4.71) with constants $I = 0.05$, $a = 0.139$, $\varepsilon = 0.008$, $\gamma = 2.5$. In the left picture, the nullclines are depicted in blue ($\dot{V} = 0$) and green ($\dot{w} = 0$). In red the orbit that starts at the origin (0.0). The trajectory quickly converges to a periodic orbit, *a stable limit cycle*. In the right picture the time series of the membrane voltage V of the orbit is drawn

4.7.4 *Izhikevich's Reduction*

A very elegant model that allows generation of different spiking and bursting behaviour was proposed by Eugene Izhikevich. The simple model has only four dimensionless parameters, a, b, c and d. The model is given by

$$\dot{v} = I + v^2 - u \tag{4.73}$$

$$\dot{u} = a(bv - u) \tag{4.74}$$

where the 'reset rule' is: if $v \geq 1$ then $v \leftarrow c, u \leftarrow u + d$. Izhikevich suggests to rewrite this to

$$C\dot{v} = k(V - V_r)(V - V_t) - u + I \tag{4.75}$$

$$\dot{u} = a\{b(V - V_r) - u\} \tag{4.76}$$

where the 'reset rule' is now slightly changed to if $V \geq V_{\text{peak}}$ then $V \leftarrow c, u \leftarrow u + d$ with V_{peak} the height of the action potential generated by the model. The membrane capacitance is C, the membrane potential is V, the recovery current is u and the resting membrane potential is V_r. By a proper choice of the parameters $a,\ b,\ c,\ d$ the model reproduces 20 of the most fundamental spiking patterns of neurons, such as tonic spiking, tonic bursting and bistability. In Exercise 4.21 you can explore this model in more detail. One of the advantages of this model is the computational efficiency: large-scale models comprised of different neuron types can be simulated with these models [60].

4.8 Summary

In this chapter we discussed planar systems of linear and nonlinear ordinary differential equations. You understand the concepts of eigenvalues and eigenvectors and can define the stability of equilibria using the relation between the trace and determinant of the characteristic equation, both for linear and nonlinear systems. You have learned some techniques to solve systems of ODEs. You understand the power of graphical analysis of planar systems. You learned about limit cycles and have a basic understanding of bifurcations. In the last part of this chapter we discussed reductions of the Hodgkin-Huxley equations to two-dimensions, allowing the tools and concepts introduced to study the dynamical behaviour of single neurons.

Problems

4.1 Compute the eigenvectors and eigenvalues for the following matrices:

$$\text{(a)} \begin{pmatrix} -4 & -2 \\ 3 & 3 \end{pmatrix} \quad \text{(b)} \begin{pmatrix} -4 & -2 \\ 2 & -4 \end{pmatrix} \quad \text{(c)} \begin{pmatrix} -4 & -2 \\ 0 & 3 \end{pmatrix}$$

4.2 Consider the equation

$$\dot{\boldsymbol{x}} = \boldsymbol{A}\boldsymbol{x}$$

for each matrix $\boldsymbol{A}_i$ with $i = 1, 2, 3$:

$$A_1 = \begin{pmatrix} -1 & 0 \\ 0 & -2 \end{pmatrix} \quad A_2 = \begin{pmatrix} 1 & 0 \\ 0 & -2 \end{pmatrix} \quad A_3 = \begin{pmatrix} -1 & 1 \\ -1 & -1 \end{pmatrix}$$

Calculate the eigenvalues. What are the characteristics of the origin?

4.3 Consider the equation

$$\dot{\boldsymbol{x}} = \begin{pmatrix} 1 & -1 \\ 1 & 3 \end{pmatrix} \boldsymbol{x}$$

Compute the eigenvectors and eigenvalues. What do you observe? Is it now straightforward to calculate the general solution?

4.4 Is there a maximum for the number of fixed points for nonlinear 2-dimensional autonomous systems? And for linear planar autonomous systems?

4.5 With $A = \begin{pmatrix} -4 & -2 \\ 3 & 3 \end{pmatrix}$ give the general solution of (4.2). Compute the values of c_1 and c_2 when $x(0) = \begin{pmatrix} 1 \\ 1 \end{pmatrix}$.

4.6 Given the planar differential equation

$$\begin{aligned} \dot{x} &= x + 2y \\ \dot{y} &= x + 2y \end{aligned}$$

(a) What are the eigenvalues of the system?
(b) What are the eigenvectors?
(c) Sketch the solution in the phase plane for a few initial conditions.

4.7 Given the planar differential equation

$$\begin{aligned} \dot{x} &= 1 + x^2 - y \\ \dot{y} &= bx - y \end{aligned}$$

where $b > 0$.

(a) For which values of b two equilibria exist?
(b) Show that in that case the one to the right is always a saddle point.
(c) When is the left equilibrium stable?

4.8 Is it possible that trajectories in the vector field in autonomous systems cross outside the fixed point?

4.9 Determine the equilibrium points and their stability of the following system:

$$\dot{x} = x + (x^2 + y^2)/2$$
$$\dot{y} = y + (x^2 + y^2)/2$$

Are periodic solutions possible? Use pplane[11] to draw the vector field.

4.10 Show that the nullclines for the predator prey problem, (4.39), are given by

$$\{y = 0, x = 0\}$$

and

$$\left\{y = \frac{\alpha}{\beta}, x = \frac{\gamma}{\delta}\right\}.$$

The first solution effectively represents the extinction of both species. If both populations are at 0, then they will continue to be so indefinitely. Note that both x and y cannot attain negative values to comply with biological reality. The second solution represents a fixed point. Around this fixed point, both populations show oscillations. The levels of population at which this equilibrium is achieved depend on the chosen values of the parameters α, β, γ , and δ. Sketch the vector field and the nullclines for $\alpha = 1,\ \beta = 0.01,\ \gamma = 0.5,\ \delta = 0.005$.

4.11 Given is the van der Pol oscillator

$$\frac{dx}{dt} = x - \frac{1}{3}x^3 - y$$
$$\frac{dy}{dt} = b(x - a)$$

where $b > 0$ and a are parameters. The van der Pol oscillator was proposed by the Dutch electrical engineer and physicist Balthasar van der Pol. The equation was also used by Fitzhugh and Nagumo in the Fitzhugh-Nagumo equations. Other applications include modeling oscillations of the vocal folds in phonation and bipolar disorders, see e.g. [31].

(a) Determine the equilibrium.

---

[11] dfield and pplane are copyrighted in the name of John C. Polking, Department of Mathematics, Rice University, and publicly available.

(b) Determine the linearization about the equilibrium and determine its type.
(c) Sketch the phase plane including the nullclines and the equilibrium when $a = b = 1$. Use the Matlab program pplane.

4.12 Show that the system

$$\dot{x} = ax - x^2$$
$$\dot{y} = -y$$

undergoes a transcritical bifurcation when $a = 0$.

4.13 Show that the system

$$\dot{V} = I + V^2 - u$$
$$\dot{u} = bV - u$$

undergoes a Hopf bifurcation when $b = \frac{1}{2} + 2I$ for $b > 1$ and $I > 0.25$. Numerically investigate whether this is a super- or a subcritical bifurcation using pplane.

4.14 A two-dimensional system is given by

$$\dot{r} = ar + r^3 - r^5$$
$$\dot{\Theta} = 1.$$

(a) Show that for values $-1/4 < a < 0$ three equilibria exist: a fixed point, an unstable limit cycle and a stable limit cycle (with radius larger than the unstable limit cycle).
(b) Show that if $a = 0$ the unstable limit cycle has disappeared while the stable limit cycle with radius $r > 0$ is still present.
(c) Which bifurcation therefore occurs at $a = 0$?
(d) What bifurcation occurs if we change the sign of the term of r^3 to negative?

4.15 *For students who like a challenge*. Show that the dynamical system

$$\dot{x} = ax - y + x(x^2 + y^2)(2 - x^2 - y^2)$$
$$\dot{y} = x + ay + y(x^2 + y^2)(2 - x^2 - y^2)$$

undergoes a subcritical Hopf bifurcation at $a = 0$. Use the transformation to polar coordinates

$$\dot{r} = ar + 2r^3 - r^5$$
$$\dot{\theta} = 1.$$

It is now possible to find limit cycles by exploring if solutions exist where $\dot{r} = 0$.

4.16 You learned that Hopf bifurcations come in both super- and subcritical varieties. Which of the two are more likely to result in catastrophic events?

4.17 Use the reduction to a 2D system of the HH-equations given by (4.67). Take the parameter values as shown in the caption of Fig. 4.18. Use Matlab and pplane.

(a) Take for I the values 0, 2, 4, 12, 20 pA. What happens to the stability of the fixed point on the left knee of the N-shaped V-nullcline?
(b) Show that there is a large periodic orbit before the equilibrium on the left knee loses its stability. For parameter values where this is the case *bistability* exists: a coexistence of a low voltage resting state and a tonic spiking state. Determine approximately for which values of I this is the case. Note that a neuron can switch between the two states if it receives an excitatory or inhibitory pulse.
(c) Is the Hopf bifurcation at this equilibrium sub- or supercritical?
(d) When I is sufficiently large, the equilibrium on the right knee becomes stable. Determine approximately the value of I when this happens. Why would this be called a *depolarization block*?

4.18 This exercise can be made using MATLAB and pplane. Take the Morris Lecar equations as given by (4.68) with parameter values S1 from Table 4.2.

(a) Gradually increase I from 80 to 120 μA/cm^2. Use as initial conditions $V_0 = -60$ mV and $w_0 = 0.2$. Determine a value for which the response is a large spike, followed by small (sub-threshold) spikes. What is the frequency of the small spikes? Compare this with the eigenvalues at the equilibrium.
(b) Show that in a very small interval in I, the small sub-threshold spikes grow to large spikes. This phenomenon is called a Canard-explosion, see http://www.scholarpedia.org/article/Canards.
(c) What is the type of bifurcation that has occurred?
(d) Is this the only bifurcation for these parameter settings?
(e) Change the values to S2 from Table 4.2 now using $\phi = 0.067$, $V_3 = 12$ mV, $V_4 = 17.4$ mV, while keeping the other values the same. Compare the n-nullclines in the two settings.
(f) With the same constants, see what happens when the current I increases from 39 to 41 μA/cm^2. What is the crucial difference to what happened in (i)? What is the type of the bifurcation?

4.19 Consider the Fitzhugh-Nagumo equations, given by (4.71). At a fixed point, $w = V/\gamma$ and $I = V/\gamma - V(V-a)(1-V) \stackrel{def}{=} h(V)$. As h is a cubic polynomial, it can have at most three zeros.

(a) Show that h has no local minima and maxima if $\gamma < \frac{3}{a^2-a+1}$
(b) Next assume that $\gamma < \frac{3}{a^2-a+1}$. Why is there always a unique fixed point and why can it only lose its stability through a Hopf bifurcation?
(c) Determine the condition where the trace at the linearisation vanishes.

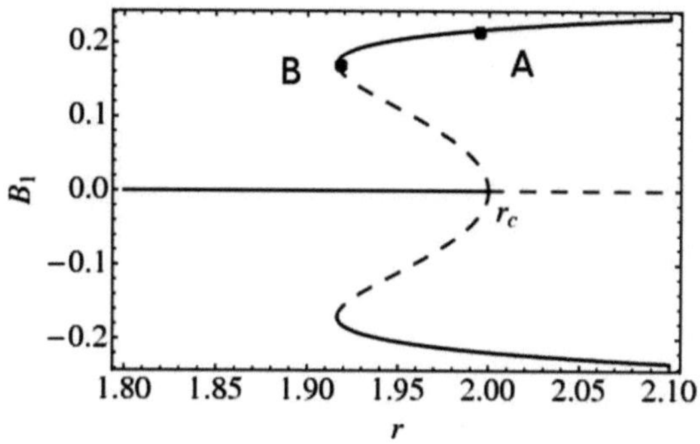

Fig. 4.21 Bifurcation diagram for the parameter r

(d) Take $a = 0.1$, $\varepsilon = 0.02$, $\gamma = 1$. Show (use MATLAB and pplane) that there is a Hopf bifurcation at $I \approx 0.54786$. Increase the applied current with very small steps and see how a periodic orbit growths rapidly. Observe that the periodic orbit initially follows the middle branch of the V nullcline. This is an example of what is called a canard. The small parameter ε in the system is responsible for this phenomenon.

4.20 Consider the bifurcation diagram shown in Fig. 4.21. Assume that the system is observed at $B_1 = 0$ and $r_0 < r_c$. If r is increased to r_c, a bifurcation occurs; let us assume that the system will arrive at A. Sketch what happens if r is decreased. What is the phenomenon called that to bring the system back to its original equilibrium the value of r needs to be smaller than r_0?

4.21 Explore the simple model from Izhikevich, as discussed in Sect. 4.7.4 using pplane and the matlabfile `Izhikevich.m`.

Part III
Networks

Chapter 5
Elementary Neural Networks and Synchrony

Pulling a good network together takes effort, sincerity and time.

— Alan Collins

Abstract All models discussed thus far were *single cell models*, but the brain has many cells that are coupled and exchange information. Connections of a few neurons can perform elementary functions, for instance filtering incoming action potential trains or detecting edges in images. More complex functions require the concerted action of neurons, where particular forms of *synchronisation* are essential for physiological function. An example is phase synchrony where neuronal assemblies synchronize the phases of their rhythms within a particular frequency band. Synchronisation is essential for information transfer between neural assemblies, ranging from odor discrimination to memory storage, execution of motor commands or performing cognitive tasks. In this chapter we present a few prototype circuits ('motifs') of interacting neurons and we discuss a few basic concepts of synchronisation.

5.1 Introduction

In previous chapters we discussed the generation of action potentials, and dynamics in one and two dimensions. We introduced two-dimensional reductions of the Hodgkin-Huxley equations, showing the rich phenomenology of neural dynamics in different conditions. We also discussed in Chap. 2 that neurons interact by chemical and electrical (gap junctions) synapses. Most neuronal functions, however, do not result from activity of single neurons, but from collective neuronal behaviour. This

Electronic supplementary material The online version of this chapter (https://doi.org/10.1007/978-3-662-61184-5_5) contains supplementary material, which is available to authorized users.

The original version of this chapter was revised: The belated corrections have been incorporated. The correction to this chapter is available at https://doi.org/10.1007/978-3-662-61184-5_11

M. J. A. M. van Putten, *Dynamics of Neural Networks*,
https://doi.org/10.1007/978-3-662-61184-5_5

takes place both within neuronal assemblies -relatively nearby collections of many neurons that form functional units- and between various neuronal assemblies that are relatively remote. For instance, assemblies involved in language production interact with assemblies involved in speech production to allow verbal communication and networks in the visual cortex connect with speech and motor areas, as well. Indeed, most functions arise by collaboration between many neurons, where sufficient *synchronization* across many spatial and temporal scales appears to be essential for information transport and function. Many neurological disorders result from dysfunction of these neuronal assemblies or their interactions. Examples include dysphasia or motor deficits resulting from a stroke, traumatic brain injury or seizures. Clinically, the dynamic interactions within and between neuronal assemblies can be studied with e.g. the electroencephalogram,[1] that we will discuss in more detail in Chaps. 6 and 7, or functional MRI.

Computational modeling can also contribute to enhance our understanding of neuronal interactions. It may identify underlying mechanisms involved in information transfer and synchronization between neurons. It may elucidate how abnormal, e.g. increased, synchronization results in seizures or tremors, and may identify potential treatment targets. Artificial neuronal networks are also used to replicate human behaviour or perception, such as driving a car, recognizing images [71] or interpretation of EEG patterns [120, 121, 131].

In this chapter, we present a few concepts to illustrate to model neuronal networks with 'individual' neurons. In Chap. 7, we discuss a complementary approach, using neural mass models: in such models, only average neuronal behaviour is considered, and details from individual neurons are left out [142]. Simulating interacting neurons may become computationally demanding. If interactions of neurons in a particular network are more important than the mechanisms involved in the generation of the action potential, neurons may be simplified. We already discussed Izhikevich' reduction in Chap. 4. An even simpler model is an integrate-and-fire neuron, introduced by the French physiologist Lapicque (1866–1952). In this model, the actual generation of the action potential is extremely simplified, which makes it computationally very fast.

5.2 Integrate-and-Fire Neurons

The *leaky* integrate and fire (LIF) model is represented by

$$\begin{cases} \tau_m \frac{dV}{dt} = E_{\text{rest}} - V + R_m I, \\ \text{if } V(t^-) = V_{\text{threshold}}, V(t^+) = V_{\text{reset}}. \end{cases} \tag{5.1}$$

Using values $\tau_m = 10$ ms, $R_m = 1$ MΩ, E_{rest} = -65 mV, $V_{\text{threshold}} = -55$ mV and $V_{\text{reset}} = -70$ mV, our LIF neuron can spike as illustrated in Fig. 5.1. The corresponding electrical circuit for the subthreshold behaviour of the LIF consists of a battery with potential V_{rest}, in series with the membrane resistance (R_m), in parallel with a

[1] Or the magnetoencephalogram, MEG.

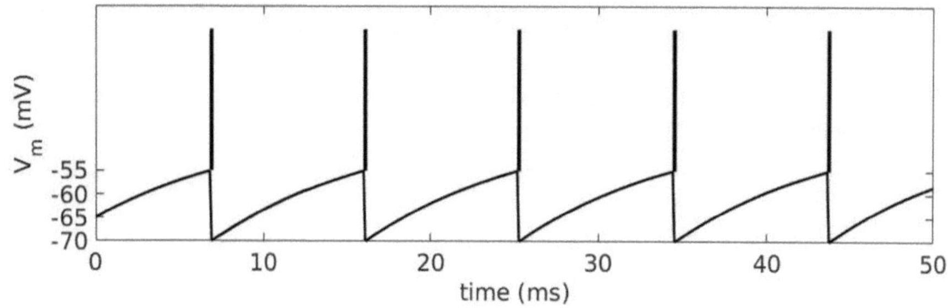

Fig. 5.1 Spiking behaviour of a LIF neuron with input current I = 20 μA. Other values as indicated in the text

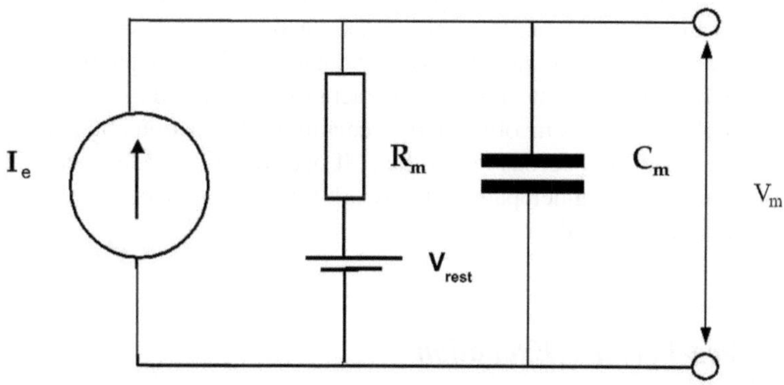

Fig. 5.2 Corresponding electrical circuit for the subthreshold behaviour of a LIF neuron

capacitor (membrane capacitance), with the understanding that if $V_m = V_{\text{threshold}}$ the circuit generates a "spike". The input current is represented by a current source. The circuit is shown in Fig. 5.2.

A variant of the LIF neuron is the quadratic integrate and fire neuron (QIF). This model is defined as

$$\begin{cases} \tau_m \frac{dV}{dt} = cV^2 + bI, \\ \text{if } V(t^-) = V_{\text{threshold}}, V(t^+) = V_{\text{reset}}. \end{cases} \tag{5.2}$$

Working with integrate-and-fire neurons is computationally cheap in comparison to neurons with voltage-gated ion channels (e.g. using the Hodgkin-Huxley model). Various educational simulators for neural circuits using integrate-and-fire neurons are available, for instance Neuronify [36], NEST (www.nest-simulator.org/) or brian (www.briansimulator.org). These simulators allow users to create and explore neural networks in a dedicated simulation environment.

Fig. 5.3 Feed-forward excitation. Neurons connect with excitatory synapses to transmit information across a particular distance

5.3 Elementary Circuits

Elementary circuits, also known as "motifs" or "micronetwork motifs", are characterized by a particular architecture between two or more neurons to realize a particular computation. For instance, elementary networks may simply propagate trains of action potentials or tune the strength and form of the efferent signals. This includes decreasing firing rates of inhibitory cells, induction of disinhibition or alterations of oscillatory coupling. Which model to use is primarily defined by the research question at hand. If network architecture and interactions are believed to be more relevant than the detailed processes involved in the generation of an action potential, models using integrate-and-fire neurons may suffice. If one wishes to understand better the effect of a particular channelopathy on network function, more complex models of a single neuron are needed.

5.3.1 Feed-Forward Excitation

Feed-forward excitation is where one neuron relays information to its neighbor (Fig. 5.3). Chains of these connections can propagate information through the nervous system. An example is the connection of a cortical pyramidal cell and an alpha motor neuron, in the anterior horn of the spinal cord. One may wonder why a synapse is needed if the primary reason is to transport axon potentials, only. Nature could also have decided to make the axon longer, perhaps?

5.3.2 Feed-Forward Inhibition

A presynaptic cell excites an inhibitory interneuron and the inhibitory interneuron then inhibits the next cell. A variant is a circuit where an excitatory neurons receives excitatory input via a direct connection, and inhibitory input via an inhibitory interneuron, that on its turn is also excited by the presynaptic excitatory neuron (see Fig. 5.4). These circuits are common in various regions of the central nervous system, including hippocampal, neocortical and thalamic networks. Feed-forward inhibitory networks can act as a low-pass filter: input to the first excitatory neuron is transmitted from the second excitatory neuron within a particular frequency range, only. This is illustrated in Fig. 5.5. Changes in feed-forward inhibitory networks have a

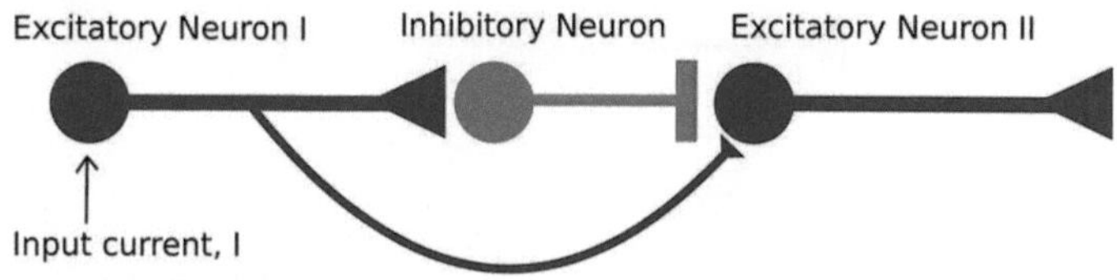

Fig. 5.4 Feed-forward inhibition. The excitatory neuron II receives both excitatory input from an excitatory neuron I and inhibitory input from an interneuron, that on its turn is excited by the primary excitatory cell, neuron I. An input current I, representing synaptic input, to the first excitatory neuron is indicated, too

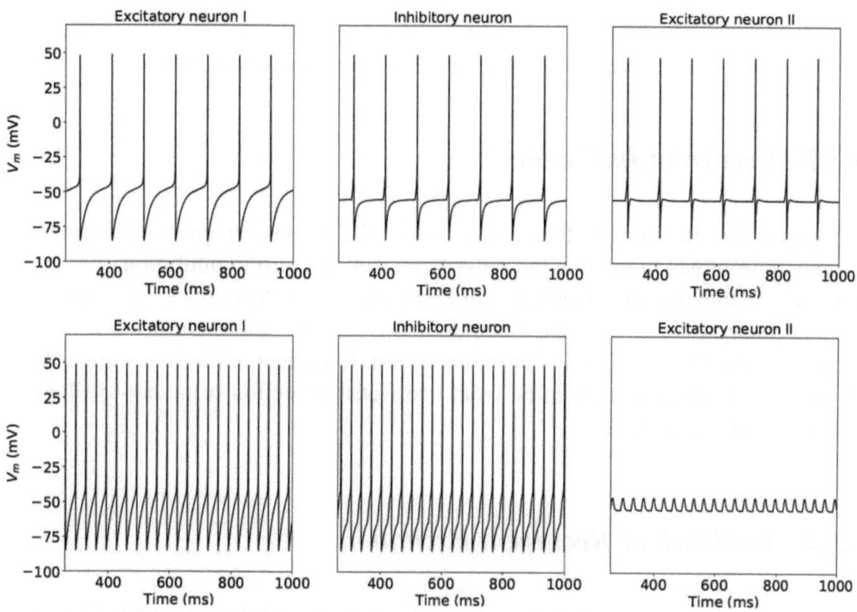

Fig. 5.5 Simulation of the effect of feedforward inhibition on spike frequency of the output neuron. The first excitatory neuron receives an input current, I. Top panel $I = 0.07$ nA; lower panels $I = 0.22$ nA. Above a particular spiking frequency, the output spiking frequency becomes zero, and only subthreshold membrane potential fluctuations remain (lower right panel). Simulations performed with synaptically coupled Hodgkin Huxley neurons in brian2

potential role in the generation of seizures, too [85, 124]. A reduction of the activity of the inhibitory interneuron by either changing the intrinsic excitability or reduction of the excitatory input has been suggested to be involved in the generation of epileptic seizures in a mouse model of Dravet syndrome [113], an uncommon but severe lifelong form of epilepsy that begins in the first year of life. Failure of feed-forward inhibition has also been put forward as a mechanism in the propagation of seizures [40] (Fig. 5.5).

Fig. 5.6 Feed-back inhibition. The receiving neuron excites an inhibitory interneuron, which inhibits the presynaptic neuron

Fig. 5.7 Feedback excitation. The receiving neuron excites the presynaptic neuron

5.3.3 Recurrent Inhibition

A presynaptic cell connects to a postsynaptic cell; the postsynaptic cell connects to an interneuron, which inhibits the presynaptic cell. Recurrent inhibition of α-motor neurons, via motor axon collaterals and Renshaw cells, reduces the spiking rate from a motor nucleus to a given synaptic input. In this way, this motif can act as a variable gain regulator at level of the lower motorneuron. Recurrent inhibition is remarkably effective: a single action potential from one Renshaw cell is sufficient to silence a motor neuron [12] (Fig. 5.6).

5.3.4 Feedback or Recurrent Excitation

A presynaptic neuron excites a postsynaptic neuron and the postsynaptic neuron excites the presynaptic neuron. This type of circuit can serve like a switch: once the presynaptic cell is activated the activation can be sustained. Such circuits play a role in e.g. the control of rhythmic activity like swimming locomotion [57]. Variants of feedback excitation exist, where a presynaptic neuron excites a postsynaptic neuron that feedbacks to excite itself (an autapse) or connects to other neurons which ultimately feedback to itself (Fig. 5.7).

5.3.5 Lateral Inhibition

A presynaptic cell excites an inhibitory interneuron that inhibits a neighboring cell in the network. This type of circuit is used in some sensory systems to provide edge enhancement. It plays a key role in the visual system where it increases contrast and sharpness. This motif is shown in Fig. 5.8.

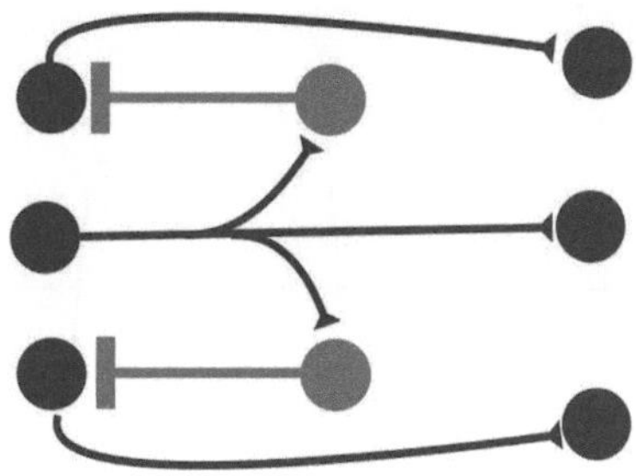

Fig. 5.8 Lateral inhibition. Can you sketch the output of the three neurons at the right given a particular input (e.g. a 3 Hz spike train) to the three neurons at the left?

5.4 Coupled Neurons and Synchrony

The elementary circuits that we discussed previously are examples of (very) small neural networks, serving a particular processing of the input. We now proceed to discuss synchronisation, loosely defined as the agreement of particular properties of two or more neurons as a function of time. Such properties include phase (phase synchronisation), amplitude (amplitude synchronisation) or phase-amplitude synchronisation. In the latter case, the phase of a particular oscillation correlates with the amplitude of a particular frequency. An example is the correlation between low-frequency EEG signals and the amplitude of bursts.

We will restrict our treatise to a few concepts only, starting with the definition of the phase of an oscillator and how the phase can respond to a particular perturbation. Thereafter, we will shortly discuss some aspects of synchronization.

5.4.1 Phase of An Oscillator

Let us consider a neuronal model with a periodic orbit with period T. For a leaky integrate-and-fire (LIF) model with a constant input current, the period T is constant, and a function of the input current. In Fig. 5.1 this was illustrated, and for that example $T \approx 10$ ms. We can then define the phase $\phi = 2\pi \cdot t/T$. In some texts, the phase is in the range 0 to 1, often preferred by biologists. If we take t as the time between two spikes, i.e. $t_1 = 0 \leq t \leq t_2 = T$, our phase is in the range $[0, 2\pi]$. Thus every point on the periodic orbit can be uniquely described with a phase ($0 \leq \phi \leq 2\pi$). Note, that the phase can be related directly to any value of the voltage in the LIF-model as the voltage is monotonous between two spikes: the voltage increases as a function of time.

For other models, the relation between the phase and the value of the voltage is different. Take for instance the periodic oscillation of a Morris-Lecar model. As the membrane voltage is not monotonously increasing (or decreasing), we cannot define the phase by considering any value of the membrane voltage, except for its maximum or minimum, and knowledge of the recovery variable w is needed as well. The phase

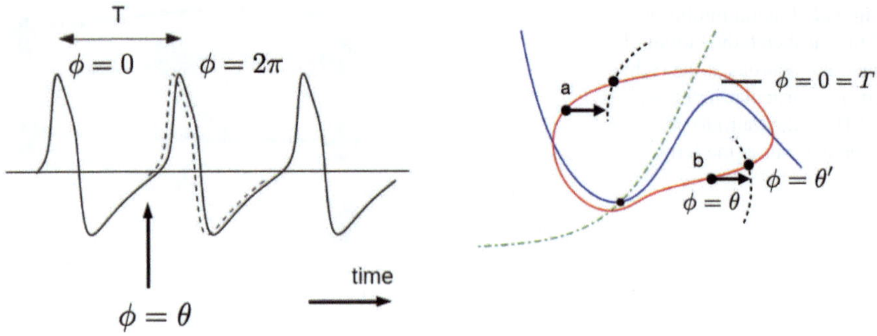

Fig. 5.9 Left: Membrane potential, V of a periodically spiking neuron. We set the phase $\phi = 0$ at the maximum value of the potential. A small perturbation is applied at $\phi = \theta$. Right: corresponding phase plane, including the nullclines (blue and dashed green) and the periodic orbit (red). Two isochrons (parts of their trajectory) are illustrated with the dashed curves. At (**a**) and (**b**) small perturbations are applied, resetting the phase backward and forward, respectively (points on the periodic orbit move anti-clockwise). Parameter values as used in Fig. 4.19

can thus be extracted from the periodic orbits of the 2D Morris-Lecar model, since a well-defined relation between the membrane voltage V and w exists.

If we would pick a point in the phase plane outside the periodic orbit, we can define its phase, too. If its trajectory ends at the periodic orbit after transients,[2] its phase is defined by the phase of the periodic orbit at that point. These curves are called *isochrons* and are typically calculated numerically. In this way, we can define the phase for any trajectory in the phase plane.

If a neuron spikes and receives additional (fast) input, for instance resulting from electrical stimulation or synaptic input, the phase will generally change. This perturbation results in an instantaneous shift of the membrane voltage, either an increase or a decrease. Input can advance the phase, resulting in an earlier spike, or set back the phase, delaying the spike. The perturbation, therefore, does not alter the dynamics, but results in the induction of phase differences. This is illustrated in Fig. 5.9.

5.4.1.1 Phase Response Curves

The magnitude of the change in phase is a function of when the perturbation is applied (at which phase) and the magnitude of the perturbation. This relationship is known as the phase resetting curve of phase response curve. A phase response curve (PRC) thus describes the transient change in the cycle period of an oscillator induced by a perturbation as a function of the phase at which it is received. First order resetting describes the change in the period containing the perturbation onset; second order resetting describes the change in the length of the next cycle (and so on). This is illustrated in Fig. 5.10. Phase resetting is an ubiquitous phenomenon in

[2]Theoretically this will take infinitely long, but numerically this is possible.

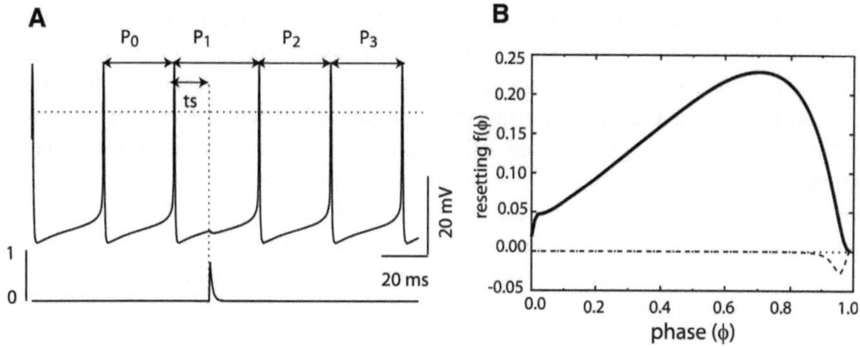

Fig. 5.10 **a**: Top trace shows the membrane potential of a spiking neuron during PRC generation. The horizontal dotted line indicates zero mV. The lower trace shows the perturbation, applied at a phase of $\phi = 0.5$. The unperturbed period is P0, the cycle containing the perturbation is P1, and subsequent cycles are P2 and P3. The stimulus time, ts, is the time between the previous spike and stimulus onset. **b** Phase resetting curve. The first order resetting is the solid line and the second order resetting is the dashed line. Third order resetting was not visible on this scale and is therefore negligible. Reprinted from [22], with permission from Elsevier Inc.

neurons [81]. It plays a role in promoting neural synchrony, for instance regulating circadian rhythms, information transmission [21], memory [111] and the regulation of cardiac rhythms via pacemaker cells. Phase resetting is also a candidate mechanism in the treatment of Parkinson's disease, essential tremor [115] and epilepsy with deep brain stimulation [86].

5.4.2 Synchronisation

Synchronisation typically refers to phase synchronisation [4, 77, 138, 141], but other types of synchrony are possible,[3] e.g. amplitude synchronisation [92]. The history of synchronization goes back to the 17th century. The Dutch scientist Christiaan Huygens reported on his observation of synchronization of two pendulum clocks which he had invented shortly before: "... *It is quite worth noting that when we suspended two clocks so constructed from two hooks imbedded in the same wooden beam, the motions of each pendulum in opposite swings were so much in agreement that they never receded the least bit from each other and the sound of each was always heard simultaneously. Further, if this agreement was disturbed by some interference, it reestablished itself in a short time. For a long time I was amazed at this unexpected result, but after a careful examination finally found that the cause of this is due to the motion of the beam, even though this is hardly perceptible.*"

[3]This is perhaps semantics, one can of course restrict synchronisation to phase synchronisation. Coupling can be used as a general term, that includes phase-phase coupling (or synchronisation), phase-amplitude coupling and amplitude-amplitude coupling [105].

Similar to pendulum clocks, neurons oscillate, too, and can synchronise their rhythms. Neurons are synchronised when their phase difference is constant. When the phase difference between two neurons $\Delta\phi = \phi_1 - \phi_2 = 0$, with ϕ_i the phase of neuron $i = 1, 2$ the synchronisation is "in-phase", when $\Delta\phi = \pi$ it is known as "anti-phase" and if $\Delta\phi$ has any other value, it is referred to as "out-of-phase" synchronisation. Let us consider two coupled, spiking neurons. We wish to explore how the spiking of one neuron affects the timing of the spiking of the other. We use as a model a LIF neuron, where the neurons are coupled via pulse coupling, i.e. the effect of the synaptic currents is instantaneous. The subthreshold behaviour of our model system is given by

$$
\begin{aligned}
C\dot{V}_1 &= g(V_{\text{rest}} - V_m + \sum_i A\delta(t - t_i)) + I \\
C\dot{V}_2 &= g(V_{\text{rest}} - V_m + \sum_j A\delta(t - t_j)) + I
\end{aligned}
\tag{5.3}
$$

with V the membrane potential, C the membrane capacitance, g the conductance, I an external current, t_i the times at which neuron 2 spikes, and t_j the spike times of neuron 1. The parameter A represents the amplitude of the perturbation, i.e. the change in membrane potential resulting from the incoming spike. Setting $A > 0$ represents excitatory coupling, negative values for A represent inhibition. In the uncoupled system ($A = 0$), both neurons behave independently. If their phases start at different values, the phase difference will remain, see Fig. 5.11, left panel. If the neurons are coupled, the spikes can quickly synchronize, see Fig. 5.11, right panel. To study more realistic scenarios, we would need several changes to our interacting neurons. This includes adding time delays to the effect of the spikes (resulting from a finite propagation velocity of the action potentials along the axons), and adding the dynamics of the synaptic transmission, as discussed in Chap. 2. It is also possible to study in more detail the effect of the strength of the perturbations or the time it takes to synchronize the neurons. This will not be treated; for additional literature, see e.g. [59].

In biological systems, including our brain, synchrony is involved in several processes, including storage of new information, movement, and recall. For many of these processes, there exists a delicate (dynamic) range in the strength of the synchrony. In some pathologies, the synchronisation is increased, as for instance during seizures or in patients with an essential tremor. A way to reduce this increased synchrony is to perturb the network with an external stimulus, as used in neurostimulation (see also Chap. 10). An example of desynchronisation of a strongly coupled network of neurons is shown in Fig. 5.12. In this simulation, a nonlinear delayed feedback signal was used to stimulate the neuronal ensemble, where the feedback signal was a function of the activity of the network [86]. Initially, the neurons are decoupled ($C = 0$) and all oscillate with slightly different natural frequencies (the rates are Gaussian distributed around the mean), and the stimulation is switched off ($K = 0$). At time $t = 300$ the coupling among the oscillators is switched on ($C = 1$) and

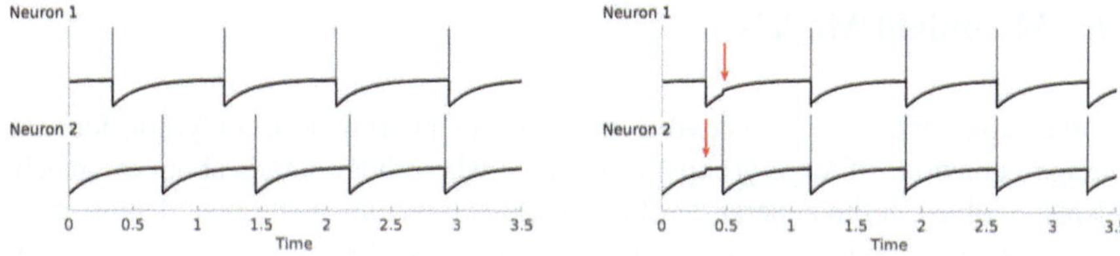

Fig. 5.11 Left: Voltage traces of two uncoupled LIF neurons and their "spikes". The initial potentials of the neurons are different, and the firing rate of Neuron 2 is slightly larger than Neuron 1. Right: Same LIF neurons, now with pulse coupling. Neuron 1 fires first and the potential of Neuron 2 increases at that time (red arrow). Similarly, when Neuron 2 fires, the potential Neuron 1 increases (red arrow). The phase shifts caused by the interaction causes the neurons to fire synchronously by the second cycle. Parameters used: $g = 5$ mS/cm^2, $V_{\text{rest}} = -65$ mV, $C = 1$ μF/cm^2; $V_{\text{reset}} = -70$ mV, $V_{\text{threshold}} = -55$ mV for neuron 2 and $V_{\text{threshold}} = -55.2$ mV for neuron 1. Each neuron receives an input current $I = 51$ μA/cm^2; the amplitude of the perturbation was set to $A = 2$ mV

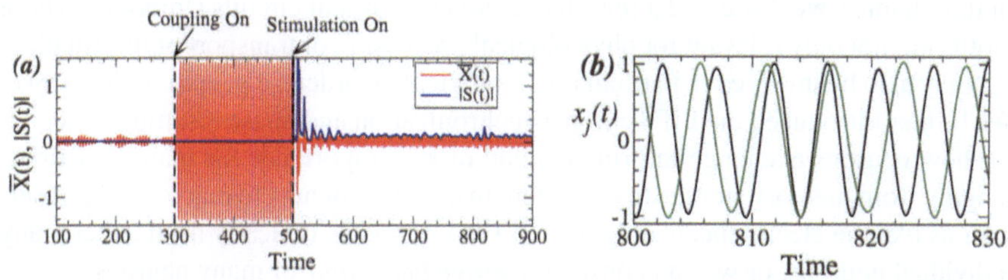

Fig. 5.12 **a** Time courses of the mean field of the ensemble, $X(t)$ coordinate, and the amplitude of the stimulation signal $|S(t)|$, red and blue curves, respectively. Coupling (C) and stimulation (K) are switched on at different times: $C = K = 0$ for $t < 300$, $C = 1$ and $K = 0$ for $t \in (300, 500)$, and $C = 1$ and $K = 150$ for $t > 500$. **b** Trajectories xj of two selected oscillators for time $t \in (800, 830)$, the stimulated regime. The desynchronisation is apparent. Illustration from [86]. Reprinted with permission from Springer Nature

the oscillators in the population become synchronized, reflected by high amplitude oscillations. At $t = 500$ the stimulation is switched on ($K = 1$) and the assembly desynchronizes.

5.5 Central Pattern Generators

An important class of networks are central pattern generators (CPG): small neural circuits (often located in the brainstem and spinal cord) of interconnected excitatory and inhibitory neurons, that generate periodic output in the absence of sensory feedback, e.g relevant for respiration, walking, swimming, or gastric motility [109]. While these networks generate the cyclic output autonomously, they are typically controlled by other networks or neuromodulatory substances in the blood. Through these input signals, the CPG can e.g. change its frequency.

5.6 Meanfield Models

Another approach to study network behaviour or neurons is to only consider the average behaviour of large groups of neurons, where both details about the mechanisms involved in the generation of action potentials and anatomical details about interactions of individual neurons are not considered. Such models are known as meanfield or neural mass models, discussed in Chap. 7. Before we turn to these latter models, we will first discuss basics of the EEG in Chap. 6. Thereafter, we turn to meanfield models of the EEG in Chap. 7.

5.7 Summary

In this chapter, we discussed a few characteristic neuronal circuits ('motifs'). These motifs are not only relevant for physiological processing or transport of information, but may also be involved in particular neurological disorders, e.g. epilepsy or stroke. We further introduced the LIF neuron, synchronization and phase resetting, essential for how neurons exchange information and of key importance for many functions, ranging from memory storage to cognition. In order to connect to macroscopic observations like the electroencephalogram (EEG) our models typically need either many individual neurons, or we can consider *average behaviour* of many neurons.

Problems

5.1 A variation of the feedback inhibition motif is the feedback circuitry where the receiving neuron excites an inhibitory interneuron, that subsequently inhibits the receiving neuron. This circuitry is present in the spinal cord where descending cortical input makes synaptic connections with the alpha motorneuron and the alpha motorneuron itself activates the Renshaw cell via alpha motorneuron collaterals.

(a) Sketch the corresponding circuitry.
(b) Perform a simulation, for instance using Neuronify[4] or NEST showing differences in the firing frequency of the alpha motorneuron with and without a functional Renshaw cell.

5.2 A feedforward inhibitory network can act as a low-pass filter. Create a model (e.g. using Neuronify or write your own Matlab code[5]) and show that this is indeed the case. Use as input a spike train with different frequencies. Note, that this very

[4]For an overview of software packages, see Appendix A.

[5]Another software package for simulating interacting integrate-and-fire neurons is NEST https://www.nest-simulator.org/.

simple network can also be viewed as a *feature detector*: it responds to low frequency input, and ignores frequencies above a particular threshold. This property was also shown in Fig. 5.5.

5.3 Central pattern generators are used in for instance chewing or walking. These generators are also present for many autonomous functions, such as bowel movement and respiration. Can you make a simple model for such a central pattern generator? For instance, a circuit with excitatory and inhibitory interneurons can realize a pattern generator for walking where the initial control signal to both limbs is initially identical.

5.4 Derive the expression for the firing frequency of a leaky integrate and fire neuron as a function of the input current, I, the membrane time constant τ, E_{rest}, $V_{threshold}$ and Vreset.

5.5 Use the Matlab program `IF_coupled.m` to study the effect of inhibitory coupling on the synchronization of the two coupled LIF neurons. If the two natural frequencies differ, is synchronization still possible?

Part IV
The Electroencephalogram

Chapter 6
Basics of the EEG

We see in the electroencephalogram a concomitant phenomenon of the continuous nerve processes which take place in the brain, exactly as the electrocardiogram represents a concomitant phenomenon of the contractions of the individual segments of the heart

— Hans Berger

Abstract In this chapter, we introduce the essentials of the generation of the EEG. We discuss current dipole sources to model the ionic currents and associated potentials generated by pyramidal cortical cells. We explain why the EEG mainly reflects synchronous activity from large assemblies of these pyramidal cells. In the second part of the chapter, we give an introduction to clinical EEG recordings and its role in ischaemia, epilepsy and coma.

6.1 Introduction

Ionic currents in the brain mainly result from synaptic transmission and the generation of action potentials. The intra- and extracellular currents create voltage differences that we can measure at various locations in the extracellular space and at the scalp. We will show that the EEG mainly results from currents generated by synchronous synaptic activity to cortical pyramidal cells. As several neurological disorders involve changes in cortical function, the EEG has been a standard tool in clinical neurology for almost a century. Applications range from disease classification or seizure detection in epilepsy, assessment of neurodegenerative disorders

---

Electronic supplementary material The online version of this chapter (https://doi.org/10.1007/978-3-662-61184-5_6) contains supplementary material, which is available to authorized users.

The original version of this chapter was revised: The belated corrections have been incorporated. The correction to this chapter is available at https://doi.org/10.1007/978-3-662-61184-5_11

M. J. A. M. van Putten, *Dynamics of Neural Networks*,
https://doi.org/10.1007/978-3-662-61184-5_6

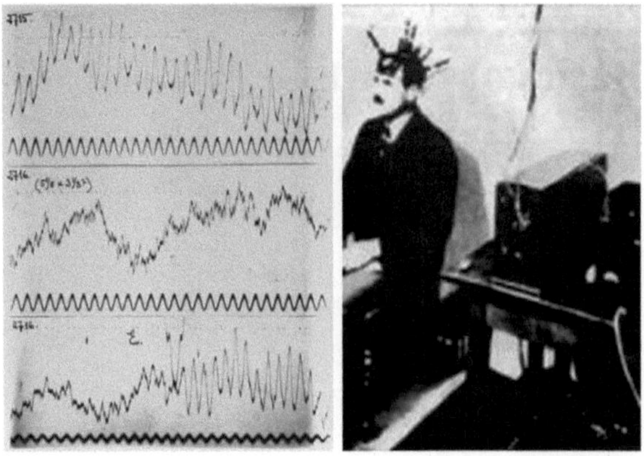

Fig. 6.1 Left: One of the first human EEG recordings. The alpha rhythm is clearly visible, including suppression by opening of the eyes (middle panel). The sine wave is a calibration signal. At the right, a detail of the string galvanometer is displayed

(e.g. dementia), and prognostication in coma to diagnostics for sleep disorders and brain monitoring in the operating theatre and the intensive care unit. The EEG is also extensively used as a pre-clinical research tool, for instance to study language processing or attention.

The first recording of the EEG through the intact skull was realized by Hans Berger in 1925. His initial measurements were performed on his son, Klaus, at that time about eleven years old. An example of such a recording is shown in Fig. 6.1. Berger used the string galvanometer for his measurements. The first rhythm he discovered was the alpha rhythm, with a frequency of 8–13 Hz. Only after 5 years and re-evaluation of his data, Berger published his results [11]. He wrote that "the brain generates electrical impulses or waves". These change if the eyes open or close. He also discovered that mental efforts, in conditions with the eyes closed, suppressed the alpha rhythm, and induced faster activity in the β range, 13–25 Hz. The publication of his findings in 1929 was a major breakthrough in neurophysiology. We will discuss in more detail how the EEG is generated, using our current understanding of neurophysiology.

6.2 Current Generators in the Brain

Transmembrane ionic currents, resulting from action potential generation, and synaptic input to neurons generate voltage differences across the cell membranes, as we discussed in Chaps. 1 and 2. At current sources, positive charges are transported from the neuron into the extracellular space, while current sinks transport positive charges into the intracellular space. For instance, inhibitory synaptic input serves as a current source, as positive charges leave the neuron. Excitatory input serves as a current sink, as positive charges enter the neuron.

As currents always form closed loops (Kirchhoff's law), these transmembrane currents also result in currents in the extracellular space. These extracellular currents generate non-zero voltages outside a region surrounding the neuron: *local field potentials* (LFPs).

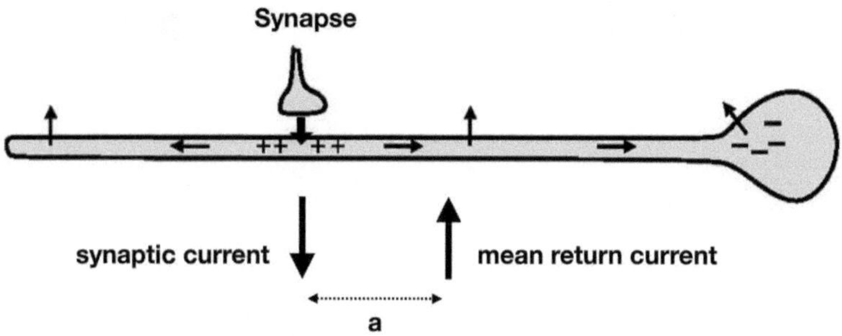

Fig. 6.2 Cartoon of currents entering and leaving a neuron. A current sink (the point of view is the extracellular space) is illustrated at the dendrite, where positive charge enters the neuron. At various locations, current leaves the dendrite. As current densities are different along the dendrite, the *mean return current* is generally located at a non-zero distance, indicated with a, from the synaptic current, resulting in a current dipole

The return current occurs at a site remote from the synapse. In pyramidal cells the return current positions are relatively far apart and non-symmetric. This causes the location of the current leaving (entering) the neuron at the synapse to be different from the mean current leaving (entering) the neuron. This is illustrated in Fig. 6.2. A non-symmetric return current implies that the *mean return current* is separated from the synaptic current along a principal axis. We can therefore assume that the mean return current is generated at a particular distance a from the synapse, effectively creating a *current dipole*.

As we will quantify in the Sect. 6.3, to record signals above noise level at a more remote location from the neurons, for instance the scalp, two other conditions need to be satisfied, too. First, a sufficiently large number of cells must be synchronously active, and second, the extracellular currents need to be more or less aligned. As groups of pyramidal cells receive, to some extent, synchronous input, the first condition is satisfied. Second, the dendrites of pyramidal cells, receiving most input, are aligned in parallel. Neurons that have a more or less symmetric distribution of dendrites, e.g. stellate cells, do not create significant voltages outside a region surrounding the neuron, as the sum of the various current dipole sources results in a zero potential. This is also known as a *closed field*. Similarly, if pyramidal cells are not aligned, the potential relatively remote from this assembly will be close to zero. If several pyramidal cells are aligned, this can result in an effective current dipole with a strength larger than the individual current dipole, and an *open field* results. This is illustrated in Fig. 6.3.

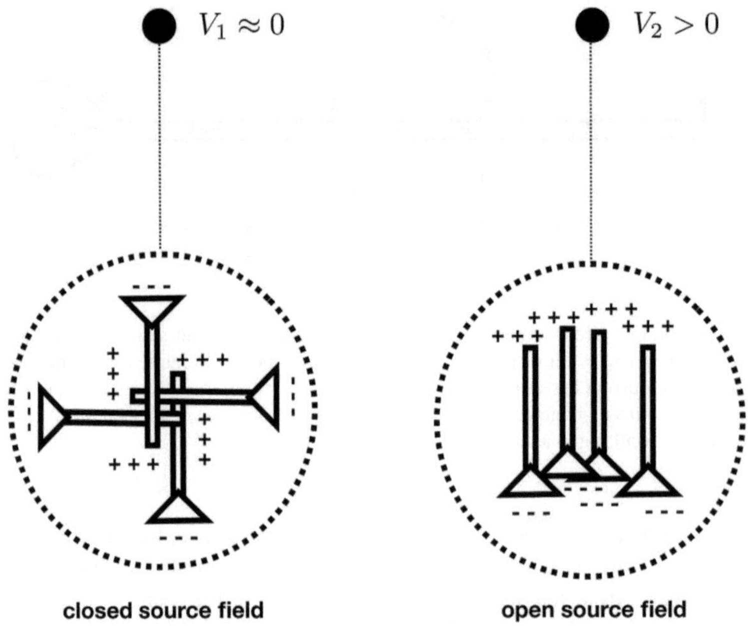

Fig. 6.3 Left: closed fields do not generate significant potential differences outside a region surrounding the neuron or neuronal assembly (indicated with the dashed circle), and at a remote position $V_1 \approx 0$. Illustration shows non-aligned pyramidal cells; one could also picture a stellate cell. Right: open source field, e.g. resulting from a group of aligned pyramidal cells. This can be represented by an equivalent current dipole. At a measurement position remote from a region surrounding the neuron or neuronal population it now holds that $V_2 > 0$

6.3 EEG and Current Dipoles

Volume-conductor theory and cable theory essentially define how much each neuron contributes to he voltages that can be measured in the extracellular space. This is illustrated in Fig. 6.4 showing a computer simulation of single synaptic input to a pyramidal cell with realistic morphology. The generated local field potentials resulting from the extracellular currents are displayed in the contour plot, and displayed as a function of time at two recording positions.

We will discuss the relation between voltages recorded at the scalp and the current dipole as a model for the currents associated with synaptic input to a pyramidal cell. Part of our discussion is strongly motivated by a treatise from Kevan Hashemi (http://www.opensourceinstruments.com/Electronics/A3019/EEG.html). Let's start with a single synaptic current I that enters (current sink, excitatory input) or leaves (current source, inhibitory input) the neuron at the synaptic density. The mean return current leaves the neuron at a distance a from the synapse. We wish to find an expression for

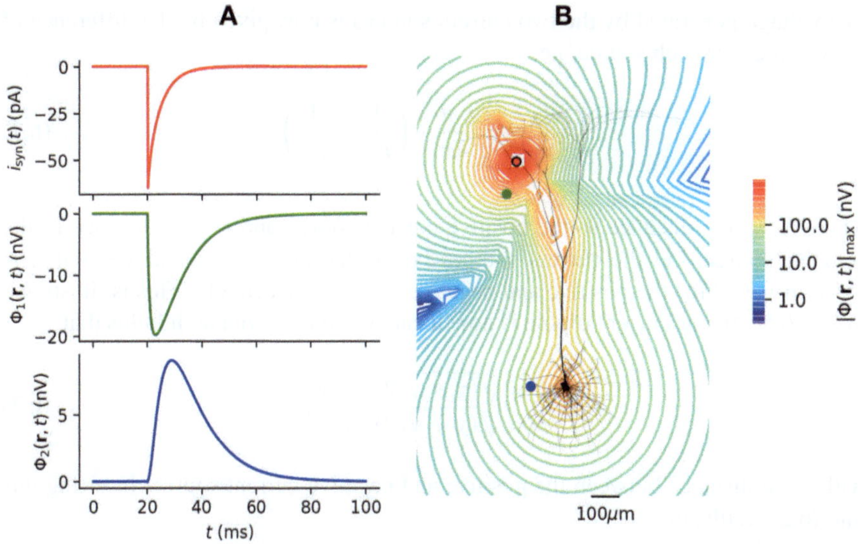

Fig. 6.4 **a**: simulated synaptic current and local field potentials as a function of time recorded in the proximity of the synapse (middle panel, green) and remote lower panel, blue) and near the soma in a passive model of a pyramidal cell. The two local field potentials at these recording positions have opposite signs, illustrating the characteristics of a current dipole. **b**: contour plot of the *absolute* value of the maximum voltages. The extracellular field potentials are of the order of several nano Volts, which is a typical value resulting from single synaptic input currents. Simulation performed with LFPy, a Python package for calculation of extracellular potentials from multicompartment neuron models and recurrent networks of multicompartment neurons. It relies on the NEURON simulator and uses the Python interface. The pyramidal cell morphology is from Mainen and Sejnowski, J Comput Neurosci, 1996

the voltage that results from a single current dipole[1] in the plane. We thus consider two current sources of equal size, but opposite signs, evaluated in the x,y-plane. Recall that the voltage generated by a single mono-polar electrode generating a current I, in a homogeneous conductor with conductivity σ is given by[2]

$$V(x, y) = \frac{I}{4\pi\sigma}\frac{1}{r}. \tag{6.1}$$

[1]Note that this is different from an electric dipole consisting of a positive and a negative *charge*, separated by a distance a, where the resulting voltage is given by $V = \frac{Qa}{4\pi\epsilon_0}\frac{x}{r^3}$.

[2]The derivation is straightforward: consider a current source with strength I. Consider a sphere around the source, with surface $4\pi r^2$. The voltage difference dV generated over the resistance surface of the sphere with thickness dr is given by $dV = I\frac{dr}{\sigma A} = I\frac{dr}{\sigma 4\pi r^2}$. Integrating results in the equation given. Remember, that the resistance of the spherical surface with thickness dr is given by the ratio of dr and the product of the conductivity and the area of the sphere.

The voltage generated by the two current sources is now given by the difference of the voltages for each source, i.e.

$$V(x, y) = \frac{I}{4\pi\sigma}\left(\frac{1}{r_+} - \frac{1}{r_-}\right) \tag{6.2}$$

where r_+ is the distance to the positive current source and r_- the distance to the negative source. Let the current sources be located at $x = a/2$ and $x = -a/2$ at position $y = 0$ for the positive and negative source, respectively. This is illustrated in Fig. 6.5. As our current dipole is now located in the x-y-plane, it holds that

$$\begin{aligned} r_+ &= \sqrt{(y - a/2)^2 + x^2} \\ r_- &= \sqrt{(y + a/2)^2 + x^2} \end{aligned} \tag{6.3}$$

with a the distance between the positive and negative current source. Inserting this into (6.2) results in

$$V(x, y) = \frac{I}{4\pi\sigma}\left(\frac{1}{(y^2 + a^2/4 - ya + x^2)^{0.5}} - \frac{1}{(y^2 + a^2/4 + ya + x^2)^{0.5}}\right) \tag{6.4}$$

Using $r^2 = x^2 + y^2$ yields

$$V(x, y) = \frac{I}{4\pi\sigma r}\left(\frac{1}{(1 - \frac{ya - a^2/4}{r^2})^{0.5}} - \frac{1}{(1 + \frac{ya - a^2/4}{r^2})^{0.5}}\right). \tag{6.5}$$

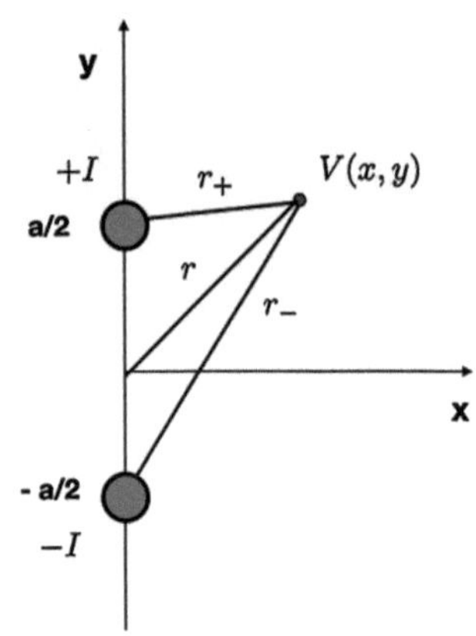

Fig. 6.5 Current dipole with dipole moment $p = I \cdot a$, with a the distance between the current sources. The voltage $V(x, y)$ is given by the difference of the voltages for each source

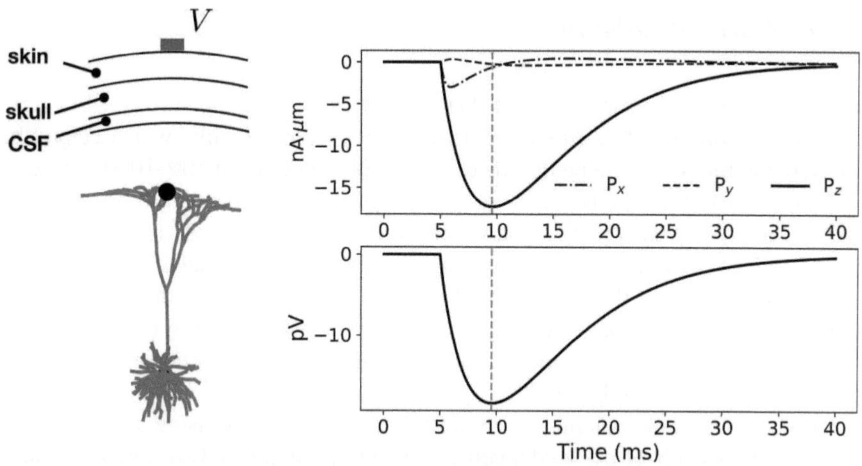

Fig. 6.6 Left: Single synaptic input to a cortical pyramidal cell in a four sphere head model, with different conductivities for brain tissue, CSF, skull, and skin. Right upper panel: Synaptic input starts at t = 5 ms and results in a maximum dipole strength of approximately 17 nA·µm. Note the three directions of the dipole strength along the x, y and z-axis, where the z-axis is in the direction of the dendrite (upwards in the plane of the figure). Right lower panel: Voltage measured at the skull. The maximum value is approximately 20 pV. Simulation with LFPy, using code provided as example4.py

Assuming that $y \gg a$ and $a \ll r$, thus $a^2/4 \ll ya$ we find

$$V(x, y) \approx \frac{\mathbf{p}}{4\pi\sigma}\frac{1}{r}\left(1 + \frac{1}{2}\frac{ya}{r^2} - 1 + \frac{1}{2}\frac{ya}{r^2}\right) = \frac{\mathbf{p}}{4\pi\sigma}\frac{y}{r^3} \tag{6.6}$$

with the current dipole moment $\mathbf{p} = I \cdot a$ with units Ampère m (A· m). At a distance far from the current dipole, along the axis of the dipole, the potential now drops with $1/r^2$ (as $y = r$). If we move away perpendicular to the dipole, i.e. in the plane where $y = 0$, $V = 0$.

We can now calculate the contribution of a single current dipole to the voltage measured at the skull. Let's assume that the dipole moment is 17 nA·µm, and that the distance from the skull to the dipole is 10 mm. If we assume that this is larger than the separation of the currents, we use (6.6) to arrive at a voltage 45 pV. A more precise estimation can be obtained using a four sphere head model, with conductances $\sigma = 0.3, 1.5, 0.015, 0.3$ S/m for brain, cerebrospinal fluid, skull and skin, respectively and corresponding radii with $r = 79$, 80, 85 and 90 mm. Using the morphology of a pyramidal cell with the same current dipole moment of 17 nA·µm, we arrive at a maximum voltage of approximately 20 pV, which is the same order of magnitude. The results are shown in Fig. 6.6.

6.3.1 Cortical Column

We showed in the previous section that the contribution to the surface EEG from a single synaptic input to a pyramidal cell is a few pico V, only, which cannot be measured at the scalp. If a neuron receives synaptic input from 500–1000 synapses, the voltage may increase to a few nano V, but this is still too small to be measured at the scalp. As we can record signals at the scalp with amplitudes ranging from 10–500 μV, a large number of cortical pyramidal cells must be activated simultaneously and the orientation of the current dipoles needs to be parallel. Both conditions are satisfied: part of the input to the dendrites of pyramidal cells is synchronized and pyramidal cells are aligned, grouped into a common structure, known as a cortical column [112], shown in Fig. 6.7.

These cortical columns may be viewed as elementary processing units, containing many thousands of pyramidal cells, glia and interneurons. Our goal is to make an estimate of the extracellular voltage measured at position $x = 0$, $y = a + h$ that results from the activation of multiple pyramidal cells contained in a particular cortical column with radius R. Contributions from other cells (interneurons, glia) to the EEG will be ignored. This is reasonable, as interneurons are not well-aligned, and essentially generate closed fields if evaluated at a distant site like the scalp. Also,

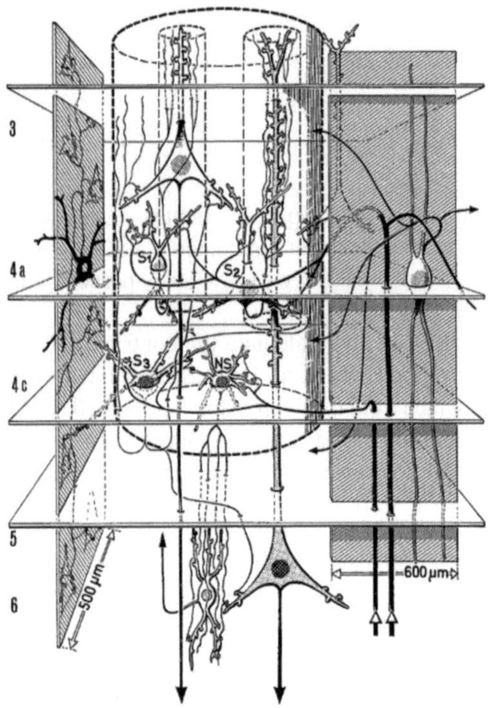

Fig. 6.7 Stereoscopic view of the elementary neuron circuit in sensory cortical areas. The main contribution to the extracellular voltage is from the pyramidal cells, as these cells create open fields and their dendrites are aligned parallel, perpendicular to the cortical surface. Superficial pyramidal cells receive input from other cortical neurons, while the deep pyramidal cells receive both intracortical and subcortical input, mainly the thalamus. The numbers 3–6 refer to the cortical layers. Reprinted from [112], with permission from Elsevier

while glia is metabolically very active, and also contributes to changes in local field potentials, the contribution to EEG is very small [20].

To estimate the scalp voltages resulting from synaptic input to pyramidal cells in a cortical column we proceed as follows. Assuming that each pyramidal cell can be modeled by a single current dipole, we consider a collection of these dipoles contained in a cortical column as illustrated in Fig. 6.8. For the voltage at the electrode due to a single neuron (current dipole) at position x we write, using $r_1 = (h^2 + x^2)^{0.5}$ and $r_2 = ((h + a)^2 + x^2)^{0.5}$:

$$V(x) = \frac{I}{4\pi\sigma}\left(\frac{1}{(h^2 + x^2)^{0.5}} - \frac{1}{((h + a)^2 + x^2)^{0.5}}\right). \tag{6.7}$$

As each neuron makes a contribution to the extracellular potential recorded at our electrode, we can add the various neurons in an annulus from $x = x$ to $x = x + dx$ to derive the voltage dV, expressed as

$$dV = 2\pi x \cdot dx \cdot V(x) \cdot S \tag{6.8}$$

with S the density of the neurons (neurons per m^2). For all neurons in a cylinder with radius R we now integrate from $x = 0$ to $x = R$, i.e.

$$\begin{aligned} V &= \textstyle\int_0^R dV = \int_0^R \frac{IS}{2\sigma}\left(\frac{x}{\sqrt{x^2+h^2}} - \frac{x}{\sqrt{(a+h)^2+x^2}}\right) dx \\ &= \tfrac{IS}{2\sigma}\left[\sqrt{(x^2 + h^2)} - \sqrt{x^2 + (a + h)^2)}\right]_0^R \\ &= \tfrac{IS}{2\sigma}\left[R\sqrt{(1 + (h/R)^2)} - R\sqrt{(1 + ((a + h)/R)^2)} + a\right]. \end{aligned} \tag{6.9}$$

This expression can be further simplified if we assume that the radius R of the column is large w.r.t the distance from recording position to the cortical column, i.e. $R^2 \gg (a + h)^2$. We then obtain

$$V \approx \frac{IS}{2\sigma}\left[a - a^2/2R - ah/R\right] = \frac{IaS}{2\sigma}\left[1 - a/2R - h/R\right]. \tag{6.10}$$

If $2R \gg a$ and $R \gg h$ this results in

$$V \approx \frac{\mathbf{p} \cdot S}{2\sigma} \tag{6.11}$$

with $\mathbf{p} = \mathbf{I} \cdot a$ the current dipole moment with unit A.m. We now use boldface notation for the dipole moment and the current to indicate that the dipole moment is a vector in the direction of the current, $\mathbf{I}$. Equation 6.11 thus expresses the voltage that results from all neurons contained in a cylinder with radius R where each neuron is represented by a current dipole with a current $+I$ at the top and $-I$ at the bottom when the recording position is relatively close to the column. Note, that in this

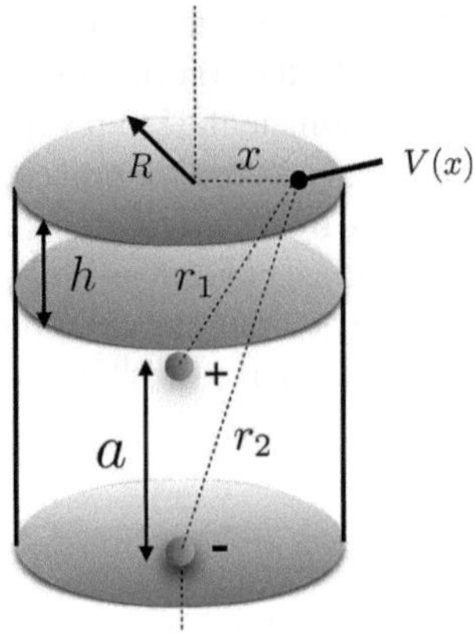

Fig. 6.8 Current dipole in a cortical column with length a. The electrode is positioned at the center of the column at height h above the column

expression *the voltage is independent of the recording distance, h, from the column!* This can be explained as follows: although the potential from the dipoles directly beneath the electrode falls of as $1/h^2$, we assumed that $R \gg h$. This implies that the contribution from distant current sources to the recorded potential is essentially constant as for large x, with $x \gg h$ it holds that $(h^2 + x^2)^{0.5} \approx x$. If the assumptions resulting in (6.11) do not hold, i.e. $h \gg R$, however, (6.9) must be used.[3] The voltage as a function of height (h) for for different column radii as a function of h is illustrated in Fig. 6.9. As an example, assume that we record from a column with radius $R \gg h$ (i.e. we can use (6.11)) and a height of the column $a = 0.5$ mm as an estimate of the length of the apical dendrite of the pyramidal cells. We can now make an estimate of the voltage generated by all these neurons, assuming a neuron density of 10^{10}m^{-2}. Using $\sigma = 0.3$ S/m and a net excitatory current of 100 pA per neuron, we arrive at $V = 1.7$ mV. If we cannot make these assumptions, and the exact solution should be used, the voltage will be *smaller*.

To summarize, the EEG reflects extracellular currents that result from synchronous synaptic input to the parallel dendrites aligned from many cortical pyramidal cells, as shown in Fig. 6.10. You may ask if it is only synaptic currents or action potentials as well that contribute to the EEG. First, the current dipole sources associated with the generation of action potentials do generally not overlap, thus essentially creating closed fields. Second, recall that the time course of changes in membrane voltage is relatively short for action potentials (1–2 ms), while synaptic currents typically last longer 10–100 ms. Therefore, the likelihood of current sources to be simultaneously

[3]In Exercise 6.4 we show that very far away from the column another simplification is possible.

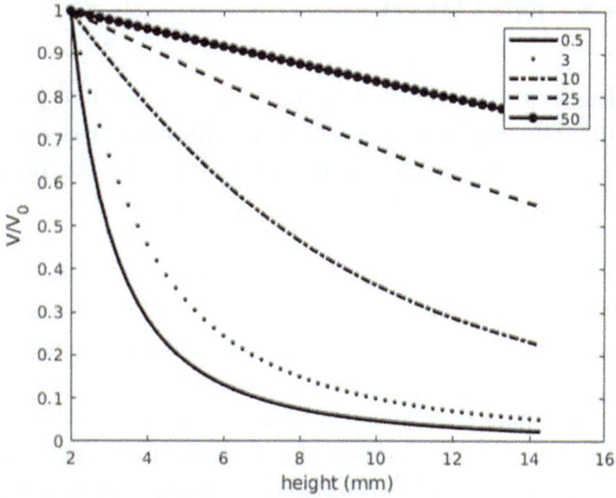

Fig. 6.9 Voltage change as a function of the distance h for different column radii $R =$ 0.5, 3, 10, 25 and 50 mm, calculated using (6.9) with neuronal density $S = 10^{10}\ \mathrm{m}^{-2}$, $\sigma = 0.3$ S/m and $a = 0.5$ mm. The larger the radius R of the dipole layer the slower the fall-off of the potential. Note that for small radii, the fall-off behaves as $1/h^2$, similar to a single current dipole (see Exercise 6.4)

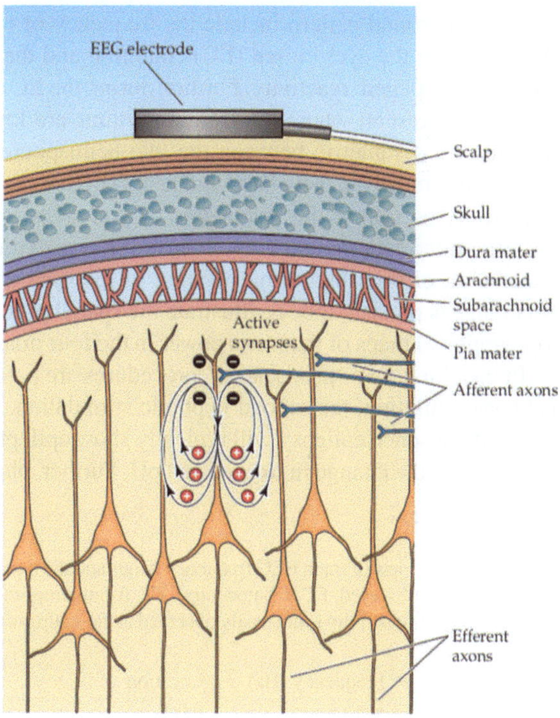

Fig. 6.10 Illustration of parallel aligned dendrites of cortical pyramidal cells receiving excitatory thalamic input. Note that the direction of the resulting current dipoles depend on where the (excitatory) thalamic input is received by the pyramidal cells. In the illustration, the afferent axons connect to the proximal dendrite; if the input were closer to the soma, the orientation of the dipole would reverse Reprinted from [42], with permission from Oxford Publishing Limited

active is much larger for those originating from synaptic currents than those generated by action potentials. Finally, the number of synaptic currents is much larger than action potentials generated by pyramidal cells (only a small fraction reaches the spike threshold). This implies that action potentials do not significantly contribute to the EEG.

6.4 EEG Rhythms

The voltages recorded at the scalp are not constant, but rhythmic in nature [11, 19]. The origin of these rhythms will be discussed in Chap. 7. Although individual neurons can generate spike trains up to 500 Hz, the frequencies of the rhythms as measured with scalp EEG range from the infraslow range (< 0.1 Hz) to gamma frequencies 35–80 Hz, and high-frequency oscillations (frequencies > 80 Hz). Most of these rhythms reflect, as discussed previously, subthreshold membrane voltage fluctuations resulting from synaptic input to cortical pyramidal cells. For the clinic, one typically evaluates EEG frequencies in the range 0.1–35 Hz, where the commonly used frequency bands are summarized in Table 6.1.

Normal EEGs have several universal characteristics, but many physiological variations exist. An illustration of two normal EEG epochs is shown in Fig. 6.11. The interpretation of the EEG is based on particular features of the background pattern(i.e. the 'global average characteristics') and the presence of 'transients'. Relevant features of the background pattern include the frequency of the posterior dominant rhythm (PDR) during the eyes closed (EC) condition and the suppression of this rhythm if the eyes are opened: reactivity. Further, during the EC condition, an anterio-posterior gradient is present, where the frontal rhythms are lower in amplitude. Also, brain rhythms are symmetric between the two hemispheres. Normal EEGs can contain slower rhythms in the theta band and faster rhythms in the beta band, where the latter is typically present in the frontal and central areas (cf. Table 6.1). In Fig. 6.11 the posterior dominant rhythm is symmetric and approximately 9 Hz, there is significant attenuation of this rhythm by opening of the eyes and some faster activity in the beta range is present, too. To illustrate the spatial gradients in the rhythms, we show topographical maps of the mean power in the four main frequency bands (Fig. 6.12).

In the clinic, two 'provocation' procedures are part of the routine EEG. The first is hyperventilation, the second is photic stimulation. During hyperventilation children with absence epilepsy will typically show epileptiform discharges, as these are provoked by the change in oxygen and pH. Further, photic driving can induce epilep-

Table 6.1 Overview of main EEG frequency bands and the typical location on the skull where these frequencies are observed. FC: frontocentral. FT: frontotemporal. PTO: posterio-temporo-occipital. EC: eyes closed. γ activity is not easily observed at the scalp, as the signal-to-noise ratio is relatively low

Rhythm	Frequency (Hz)	Location	Modulation/Occurrence
α	8–13	PTO	EC, attention
μ	8–13	Central region/motor cortex	Movement
β	15–25	FC	
θ	4–8	FT	
δ	0.1–4	Diffuse	During sleep
γ	25–45	Variable	

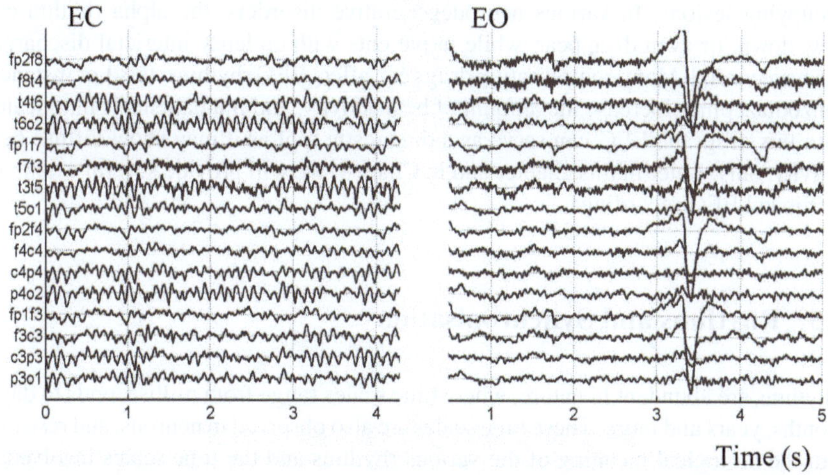

Fig. 6.11 Normal EEG with eyes closed (EC, left) and open (EO, right). The suppression of the posterior dominant rhythm is known as "reactivity". Recording using an anterio-posterior bipolar montage. Filter settings 0.5–35 Hz. The large deflection during the EO condition results from an eye blink. The letters and numbers at the left relate to the electrode position on the scalp (discussed in Sect. 6.6)

Fig. 6.12 Topographical maps of the mean power of the EEG shown in Fig. 6.11, eyes closed condition, in the four standard frequency bands. Note the bilateral occipital prominence of the alpha rhythm. Colorbars with units μV^2

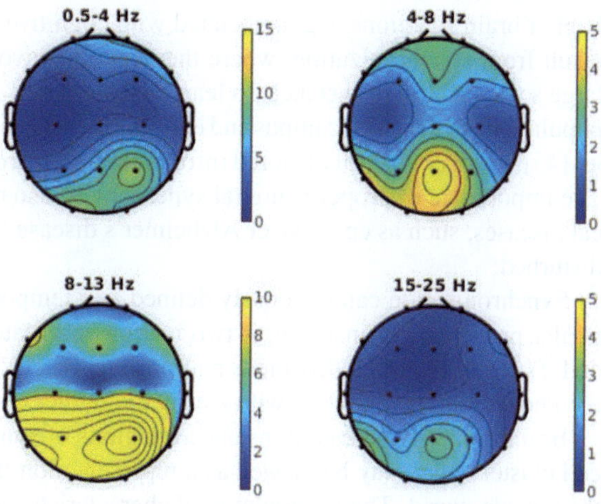

tiform discharges in children with photosensitive epilepsy, a particular generalized epilepsy, where epileptiform discharges and sometimes even seizures can be induced by exposure to rhythmic stimulation with bright light.

The frequency, location and amplitude of EEG rhythms correlate strongly to physiological and pathological brain function. For instance, during deep sleep delta activity is prominent, while this is absent during wakefulness in healthy adults. Focal delta activity can be observed in patients with cortical ischaemia or space occupying lesions. In various neurodegenerative disorders, the alpha rhythm can slow down, or even disappear, while in patients with epilepsy interictal discharges can be recorded. Many central acting drugs can affect EEG rhythms, too. For instance, benzodiazepines increase the amount of beta activity, and if a patients is intoxicated with this drug, the EEG can serve as a diagnostic tool as it may show diffuse beta activity, only. Later in this chapter and in Chap. 8, we will provide several examples of clinical EEG recordings.

6.5 Rhythms and Synchronisation

Rhythms are abundant in nature, where time scales range from milliseconds to days, months, years and more. These time scales are also observed in neurons, and research into the biological meaning of the various rhythms and the time scales involved is ongoing. In the next chapter, we will present a mean field model that simulates the various rhythms of the EEG, that result from interactions between different neuronal populations. What will not be treated in that particular chapter, however, is the presumed function of these rhythms. We will discuss some of these aspects.[4] Various basic processes, like respiration or walking, are generated by synchrony within a particular neuronal network or neural ensemble, as we also discussed in Chap. 5. Higher brain functions, e.g. associated with cognitive and executive processes, also result from synchronization, where the networks involved are typically medium to large scale [102]. For instance, to learn and store new information, several neuronal populations in the hippocampus and entorhinal cortex need to synchronise their activity [45]. For the retrieval of stored information, synchronous activity is essential, too. The importance of proper neuronal synchrony is also reflected in various neurological diseases, such as epilepsy or Alzheimer's disease [106], where this synchrony is disturbed.

Synchronisation can be loosely defined as a temporal correlation between a particular property. For instance, if two neurons oscillate at a particular frequency f_1 and f_2 we may define that these neurons are synchronized during a period T if it then holds that $f_1 - f_2 < \epsilon$ with ϵ a small number close to zero.

Brains appear to create transient formations of synchronised networks, dynamic cell clusters, that may be viewed as a representation that correlates with features of the outside world. The spatiotemporal characteristics of these dynamic cell clusters is highly variable. For instance, relatively localized clusters are found at low level stages[5] of processing of sensory input, including the retinotopic organization of the visual cortex or the somatosensory cortex. More distributed networks are associated with e.g. memory formation and conscious perception. These neural assemblies arise

[4]A great book that discusses many aspects of brain rhythms is "Rhythms of the Brain" by Buzsáki [19].

[5]Here we mean with low level stages initial routes to sensory perception, where these steps in the neural cascade are not associated with conscious perception of the input.

from transiently linking various neurons by reciprocal dynamic connections [100, 138]. These reciprocal connections may exist within the same cortical area or between different cortical regions. The dynamic formation of these assemblies results in the transient occurrence of neural assemblies, typically with survival times of the order 100-300 ms.

Various mechanisms have been proposed to be involved in the transient 'linking' of neurons [130], sometimes referred to as neural integration. An important candidate is phase synchrony, where a neural assembly result from temporarily phase locking of rhythmic activity from neuronal populations [125, 138]. Assuming two elementary populations, displaying rhythmic activity with phases ϕ_x and ϕ_y respectively, phase locking exists if

$$\Phi_{xy} = |n\phi_x(t) - m\phi_y(t)| = \text{constant} \tag{6.12}$$

for a particular duration. Here, n and m are integers that indicate the ratios of possible frequency locking.

There is a wealth of experimental evidence that phase synchronization is an important mechanism for neural synchrony, ranging from single unit and local field potential recordings to EEG measurements, both in animals and man. Other techniques that are used to identify synchronous activity in networks include functional MRI and PET. These techniques, however, suffer from a much lower temporal resolution.

In addition to phase-phase coupling, rhythms can also interact with each other by phase-amplitude and amplitude-amplitude coupling. For instance, phase-amplitude coupling has been reported between ultraslow oscillations (0.025–0.1 Hz) and the alpha and beta rhythms [105]. Also, hippocampal "ripples" (140–200 Hz) are phase modulated by sleep spindles (12–16 Hz) [63]. In pathology, this phase-amplitude modulation has been observed in patients with a postanoxic encephalopathy, where bursts may occur in clusters, where the clusters appear at defined phases of the infraslow activity [133].

6.6 Recording of the EEG

To measure voltage differences at the scalp, we need at least two electrodes. In the clinic, a 19-channel EEG recording is routine, where the electrodes are positioned as indicated in Fig. 6.13.

6.6.1 Polarity

If we measure voltage differences at the scalp, and plot the voltage as a function of time, using electrodes A and B, the convention is that if the polarity of A is more negative than B, the voltage deflects in the direction of A, and vice verse. This is

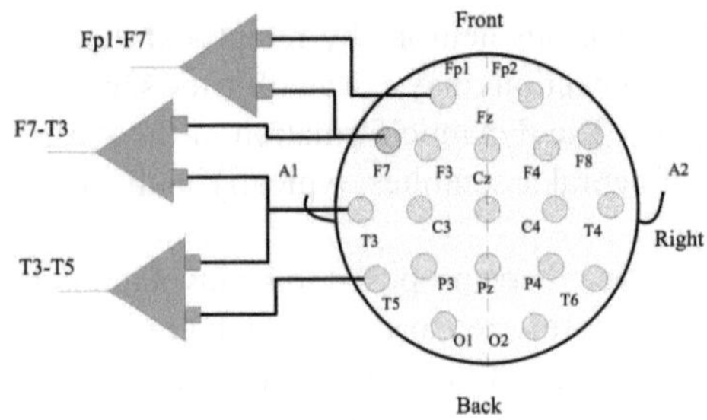

Fig. 6.13 The 10–20 system for recording the EEG. Even numbers refer to the right side of the brain, odd numbers to the left. Part of a bipolar montage is shown

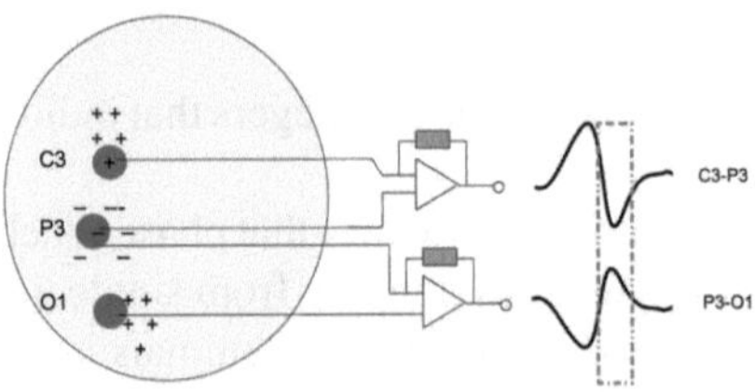

Fig. 6.14 Hypothetical charge accumulation at three electrode positions and associated voltage tracings. When P3 becomes more negative than C3 and O1, the deflection at the trace labeled C3-P3 is downwards, while in the trace labeled P3-O1 the deflection is upwards, again in the direction of P3 ("upwards"). The two curves, therefore, display a *negative phase opposition* near P3

relevant if we wish to localize the 'generators' involved in the main contribution to a particular EEG transient, as illustrated in Fig. 6.14.

6.6.2 Montages

A montage is a particular combination of recording positions (electrodes) to measure voltage differences. Commonly used montages are bipolar, e.g. from the anterior to the posterior side, transversal (from the sides of the head towards the central regions) or the use of a 'common average' where the voltages recorded are referenced to the average value of all electrodes. The reason for the use of different montages is that one actually evaluates a three dimensional current flow (cf. Fig. 6.6). As a particular bipolar or the common average montage is most sensitive to one of the current directions, in clinical assessment of the EEG it is common to use several bipolar montages for the final interpretation. In addition to these montages, a common average montage is often used.

Many neurophysiologists also use the "surface Laplacian" (or "source" montage) in reviewing the EEG. This montage essentially reflects the *local current density* at the recording position, thus providing information about the underlying (current)

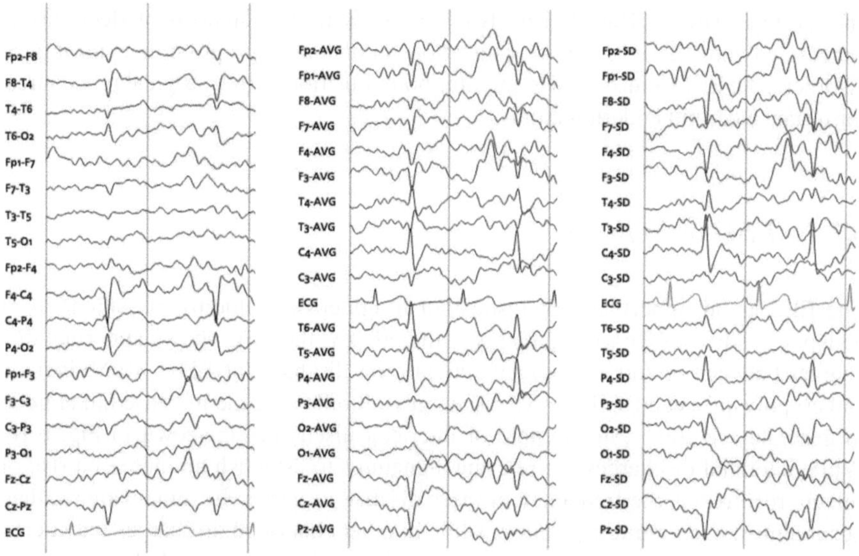

Fig. 6.15 Three EEG epochs containing the same epileptiform discharges from a 10-year old patient with Rolandic epilepsy, but in three different montages. Left: bipolar, anterio-posterior, montage. Middle: common average. Right: source montage or surface Laplacian. The unilateral occurrence of the spikes with phase opposition is best appreciated in the bipolar, anterio-posterior montage (left panel). In the common average and surface Laplacian montage, the largest peak is at C4, pointing upwards 'towards' C4, reflecting the largest negative potential at this electrode position, characteristic for a Rolandic spike

source. In the literature, this is also known as the 'Laplacian' or 'current source density' (CSD). We can approximate the Laplacian, $\mathrm{Lap}_S(V)$, at a particular recording position (i, j) using a finite difference approximation, that results in

$$\mathrm{Lap}_S(V)|_{i,j} \approx \frac{V_{i-1,j} + V_{i+1,j} + V_{i,j-1} + V_{i,j+1} - 4V_{i,j}}{h^2} \tag{6.13}$$

where h is the distance between the electrodes. The proof is provided in the appendix. In Fig. 6.15 we show the effect of the choice of the montage for a part of a routine 21-channel recording containing epileptiform discharges in a patient with Rolandic epilepsy.

6.7 Clinical Applications

Since the beginning of the previous century, the EEG has been a standard tool in clinical neurology, as many neurological diseases of the central nervous system correlate with changes in the EEG. Examples include neurodegenerative disorders, epilepsy and (cortical) stroke. The EEG is also very useful in the diagnosis of sleep disturbances and coma. In addition to applications in patients with possible neurological disorders, the EEG is also used to monitor brain function during surgical procedures, for instance carotid endarterectomy.

6.7.1 EEG in Epilepsy

The EEG is an important diagnostic tool in epilepsy, including classification of epilepsy syndromes. In most patients with epilepsy, the EEG shows abnormal discharges between seizures: interictal epileptiform discharges (IEDs). These include spikes, polyspikes and spike-wave discharges, all reflecting the increased likelihood to generate seizures.[6] An example of interictal discharges is shown in Fig. 6.16. While interictal discharges are relevant signatures to establish an increased risk of seizure recurrence, the frequency of these discharges varies. In some patients, they occur almost every minute or more, while in others interictal discharges may occur only once per 24 h. This motivates the clinical use of long-term ambulatory recordings in patients who may have epilepsy as this substantially increases the likelihood of detecting interictal discharges.[7]

Depending on the characteristics of the epileptiform transients and their location, in combination with the clinical features, it is often possible to categorize a particular epilepsy syndrome. This may be complemented by an analysis of the 'dipole strength and orientation' of the epileptiform discharge. As the generation of an interictal discharge results from a large number of pathologically synchronized pyramidal neurons, this assembly can be modeled as a single current dipole oriented in the same direction as the pyramidal neurons. The orientation and localisation of the dipole thus contains information about the location of the neuronal assembly involved in the generation of the discharge, and has diagnostic relevance. For instance, in patients with Rolandic epilepsy, the interictal discharge originates from the centro-temporal area, and the dipole location is tangential to the skull, illustrated in Fig. 6.17. The characteristics of the epileptiform discharges, the clinical symptoms and the findings from structural imaging (MRI) can be used for the classification of the epilepsy. This is relevant for both prognostication and therapeutic advice. Depending on the frequency and severity of seizures, anti-epileptic drugs are typically started as the first line of treatment. We will discuss more aspects of epilepsy in Chap. 9.

---

[6]Epilepsy is essentially a brain condition characterized by an increased likelihood of seizures [47].

[7]Transcranial magnetic stimulation (TMS) is also explored to establish a change in the presumed *cortical excitability* in patients with epilepsy. With this technique, it is possible to 'perturb' the cortex with evaluation of motor or transcranial evoked potentials [49]. See also Chap. 10.

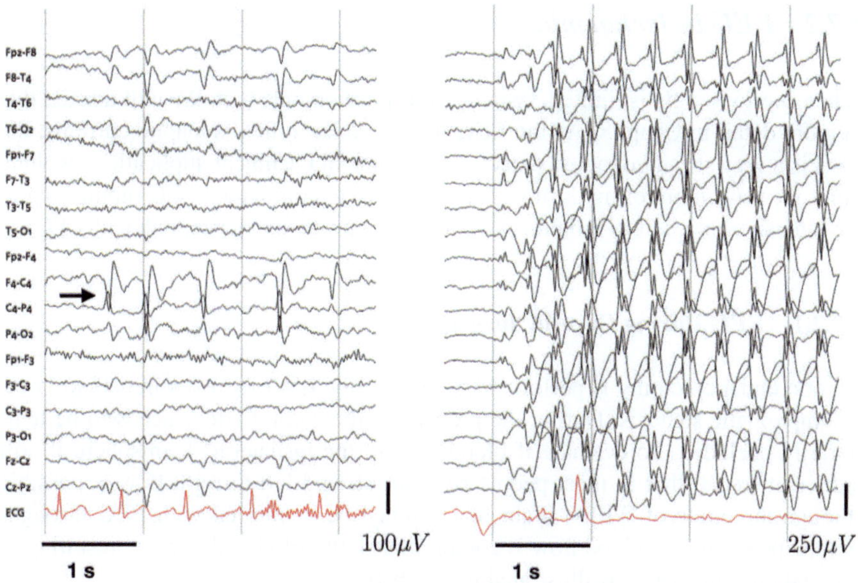

Fig. 6.16 Left panel: Interictal discharges in a 10-year old patient with Rolandic epilepsy, showing spikes over the right centrotemporal area. Note the *negative phase opposition* over C4, indicated with the arrow: the deflection from the trace recorded between F4-C4 is downwards, which implies that C4 is more negative than F4, while the trace from C4-P4 deflects upwards, which means that C4 is more positive than P4. This negative phase opposition is also observed at T4. Right panel: generalized spike-wave discharges in a 6-year old patient with absence epilepsy

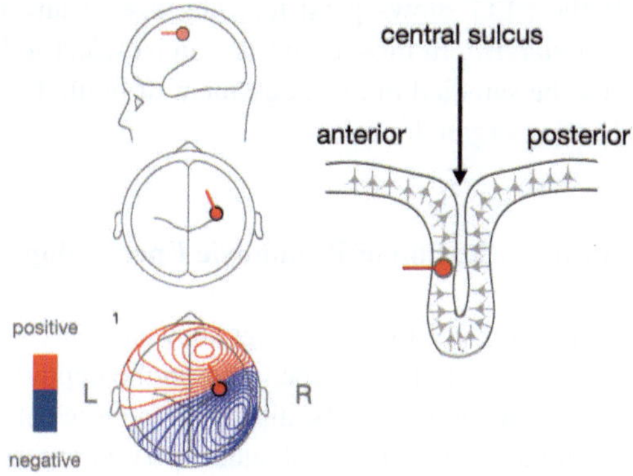

Fig. 6.17 The average current dipole of Rolandic spikes from the right centrotemporal area. The dipole is located in the primary motor strip, just anterior to the central sulcus (aka the Rolandic fissure), illustrated at the right. The pyramidal cells are all oriented perpendicularly to the cortical surface. Compare with the EEG shown in Fig. 6.16, left panel. Current dipole and voltage distribution simulated with BESA simulator. www.besa.de

6.7.2 *EEG in Ischaemia*

When energy supply is limited, synaptic transmission fails first. We will discuss this in more detail in Chap. 8. As the EEG mainly reflects these synaptic currents, the EEG is a very sensitive tool to detect acute ischaemia. Clinical applications include intra-operative monitoring during carotid surgery to continuous EEG monitoring in the intensive care unit to monitor patients with a postanoxic encephalopathy [91].

6.7.2.1 EEG During Carotid Surgery

Carotid endarterectomy is performed to prevent the risk of strokes in patients with a symptomatic stenosis. During the procedure, the carotid artery needs to be temporarily occluded. To evaluate if blood flow to the ipsilateral[8] hemisphere is still sufficient (recall that in most people right-to-left and left-to-right flow is possible through the circle of Willis) brain function can be monitored with EEG throughout the procedure. If blood flow is compromised, the EEG will change, resulting in left-right asymmetries, as illustrated in Fig. 6.18.

6.7.3 *EEG in Coma*

There is a strong correlation between changes in the EEG rhythms and changes in consciousness. If the EEG shows persistent absence of any brain activity, the recording is called *iso-electric*. In most countries, an iso-electric EEG is part of the conditions that need to be satisfied in the assessment of brain death in patients who are potential heart-beating organ donors.

6.7.3.1 EEG for Prognostication of Postanoxic Encephalopahty

Immediately after cardiac arrest, EEG recordings show no signs of electrical activity, as all synaptic activity ceased. Further, depending on the depth and duration of the hypoxia, additional neuronal damage, to be discussed in more detail in Chap. 8, may be present. Potential for recovery can be evaluated with continuous EEG recordings, as it essentially reflects synaptic recovery. This temporal evolution is illustrated in Fig. 6.19.

[8]ipsilateral: the same side as the carotid artery that is being operated.

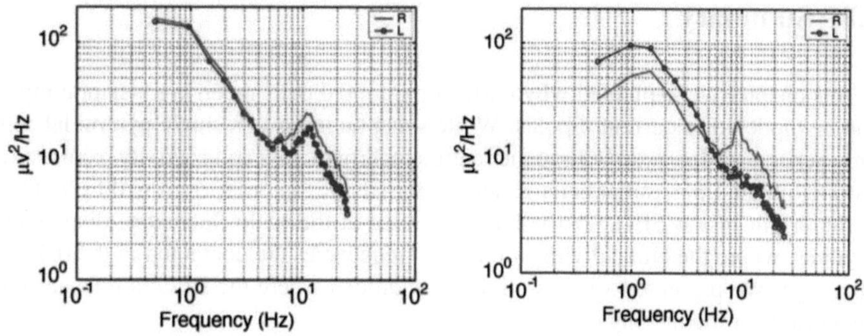

Fig. 6.18 Changes in spectral characteristics of the EEG during clamping of the left internal carotid artery. Left: Patient A shows a decrease of faster activity, only. The cerebral blood flow (CBF) in the left hemisphere is approximately 25 to 35 mL100 g^{-1}min^{-1}. Right: in patient B, faster activity is decreased with an increase in slow activity. The CBF is estimated to be approximately 15 mL100 g^{-1}min^{-1}. In both situations, shunting was advised. Modified from (van Putten et al. 2004)

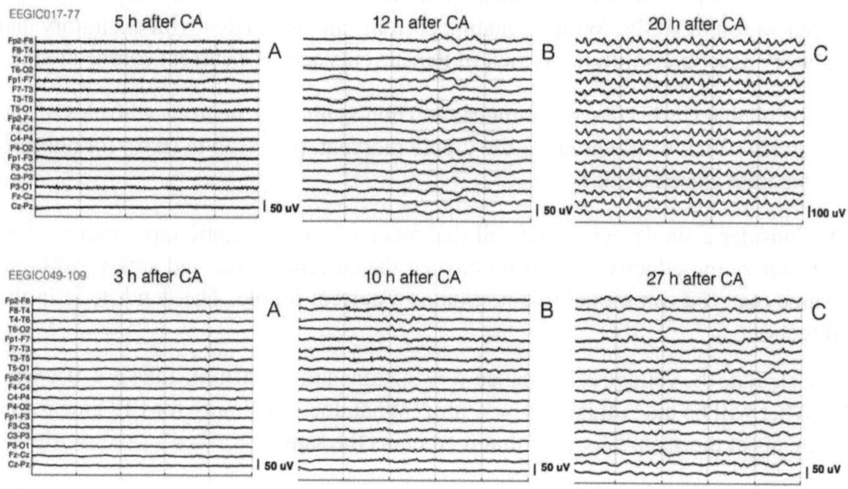

Fig. 6.19 Evolution of EEG (5-second epochs) in two patients with a postanoxic encephalopathy. Top: Favorable development, patient discharged with cerebral performance category (CPC) of 2. Bottom: Unfavorable evolution; in this patient, treatment was discontinued after 7 days. Note that even if EEG recordings in the first few hours show no signs of cerebral activity, synaptic function can recover, and prognosis can be favorable (top panels). Illustration from [91]. Reprinted with permission from Elsevier Ireland Ltd.

6.8 Summary

In this chapter we explained why cortical pyramidal cells receiving synaptic input can be modeled as current dipoles. While synaptic input to a single pyramidal cell only generates a few pico Volts at the scalp, synchronous synaptic activity from many parallel aligned pyramidal cells generates signals of the order of 10–100 μV. This is the origin of the EEG. The voltage recorded at the scalp fluctuates, displaying rhythms, where the clinically most relevant frequencies are contained in the 0.5–35 Hz range. We presented various clinical conditions where the EEG is relevant for diagnostics or monitoring.

Problems

6.1 The main excitatory neurotransmitter is glutamate, responsible for EPSPs, whereas GABA is the most abundant inhibitory neurotransmitter, responsible for the generation of IPSPs. Which synaptic currents are associated with excitatory and inhibitory neurotransmitters? Which of the currents acts as a current source or sink?

6.2 Recall that we discussed symmetric and non-symmetric neurons with closed and open fields. What would hold for the dipole moment, $\mathbf{p} = I \cdot a$ in (6.6), for neurons that generate closed fields?

6.3 Consider a single pyramidal cell that receives a net synaptic input current $I =$ 2 nA, where the effective distance between the current source and sink is 500 μm. Assume that you can model this neuron as a current dipole. The dendrite is in the y-direction.

a. Calculate the voltage at a distance $y = 0.5$ mm and $x = 0$ mm. Take $\sigma = 0.3$ S/m.
b. What will be the voltage if $y = 5$ mm? What does this imply for EEG measurements at the scalp? Assume that σ remains the same.

6.4 If you were to record very far away from the cortical column shown in Fig. 6.8 with $h \gg R$ and $h \gg a$, it can be argued that the column "behaves" like a single (averaged) current dipole. In that case, the voltage should drop as $1/h^2$ (see also Fig. 6.9). It can indeed be shown that in that situation (6.9) reduces to

$$V \approx \frac{IaS}{4\sigma} \frac{R^2}{h^2} \tag{6.14}$$

Can you prove this?

6.5 Consider two cortical columns, where one cortical column receives an excitatory current of 100 pA per neuron and the other column receives 50 pA per neuron. What is the voltage difference between the two columns? Use (6.11) $\sigma = 0.3$ S/m.

6.6 In recording the EEG, a (signal) ground electrode is used. Why?

6.7 What is the sensitivity of a routine EEG to detect interictal discharges? Can you propose techniques to increase the sensitivity?

6.8 In some forms of epilepsy, particular anti-epileptic drugs increase the likelihood of seizures, and are therefore contra-indicated. Do you know in which epilepsy type this is possible? Which anti-epileptic drugs should not be given? Can you argue why this is possible?

6.9 Why do faster rhythms (8–25 Hz) disappear first in conditions of limited energy supply?

Appendix: Derivation of the Surface Laplacian

Recall that the divergence of the gradient of the voltage V is defined as the Laplacian of V. If there are no sources or sinks within a particular region were V is measured, this is expressed in Laplace's equation as

$$\nabla^2 V = 0. \tag{6.15}$$

that in Cartesian coordinates can be written as

$$\frac{\partial^2 V}{\partial x^2} + \frac{\partial^2 V}{\partial y^2} + \frac{\partial^2 V}{\partial z^2} = 0. \tag{6.16}$$

By taking a coordinate system such that the scalp is on the x-y-plane, and using $E = -\nabla V$, we can rewrite (6.16) as

$$\frac{\partial^2 V}{\partial x^2} + \frac{\partial^2 V}{\partial y^2} = \frac{\partial E}{\partial z}. \tag{6.17}$$

As in a conducting medium it holds, using Ohm's law, that $\mathbf{E} = \rho \mathbf{j}$ with $\rho = 1/\sigma$ the resistivity (or inverse of the conductivity σ) and $\mathbf{j}$ the current density it follows that

$$\frac{\partial^2 V}{\partial x^2} + \frac{\partial^2 V}{\partial y^2} = \rho \frac{\partial j_z}{\partial z}. \tag{6.18}$$

The left hand side of (6.18) is now defined as the *surface Laplacian* of V

$$\text{Lap}_S(V) = \frac{\partial^2 V}{\partial x^2} + \frac{\partial^2 V}{\partial y^2}. \tag{6.19}$$

If $\text{Lap}_S(V)$ is nonzero, and assuming no sources are present at the recording position, current lines are diverging below the scalp, which implies the presence of a current source inside the skull.

Assume a univariate function $V(x)$, and take the Taylor series around a point $x = a$. This results in

$$V(a+h) = V(a) + V'(a)h + \frac{1}{2!}V''(a)h^2 + \frac{1}{3!}V'''(a)h^3 + \cdots . \tag{6.20}$$

Similar for h replaced with $-h$

$$V(a-h) = V(a) - V'(a)h + \frac{1}{2!}V''(a)h^2 - \frac{1}{3!}V'''(a)h^3 + \cdots . \tag{6.21}$$

Adding these two expressions, we obtain

$$V(a+h) + V(a-h) = 2V(a) + V''(a)h^2 + \frac{1}{12}V''''(a)h^4 + \cdots \tag{6.22}$$

Rewriting results in

$$V''(a) = \frac{V(a+h) + V(a-h) = 2V(a)}{h^2} - \frac{1}{12}V''''(a)h^2 - \cdots \tag{6.23}$$

If we may assume that h is sufficiently small, we arrive at

$$V''(a) \approx \frac{V(a+h) + V(a-h) - 2V(a)}{h^2}. \tag{6.24}$$

Many other possibilities exist to estimate the surface Laplacian of the EEG, including analytical differentiation after first building continuous functions from the recorded data.[9]

[9]See for further details for instance Carvalhaes and Barros: The surface Laplacian technique in EEG: theory and methods arXiv: 1406.0458v2, 2014. Parts of this section were also strongly motivated by their treatise of the surface Laplacian

Chapter 7
Neural Mass Modeling of the EEG

I believe the best test of a model is how well can the modeller answer the questions, 'What do you know now that you did not know before?' and 'How can you find out if it is true?'

— James Bower

Abstract This chapter discusses neural mass models and EEG rhythms. We start with a simple model, adding additional components step-by-step, eventually resulting in coupled neural masses that simulate a physiological EEG rhythm with a peak frequency in the 8–13 Hz (α) frequency range. While this chapter focuses on physiology, we will learn in later chapters that neural mass models also find applications in furthering our understanding of pathological EEG patterns as present during seizures or anoxia.

7.1 Introduction

The EEG is an important tool in clinical neurology and clinical neurophysiology, where applications range from diagnostics to monitoring in the ICU or operating theatre. It is also intensively used as a readout for neural function in basic neuroscience as well, e.g. to study attention, memory or language. In this chapter, we will discuss a neural mass model for the EEG generation.[1]

[1]This chapter is written by Rikkert Hindriks, with minor modifications by Bas-Jan Zandt, Annemijn Jonkman and Michel van Putten.

Electronic supplementary material The online version of this chapter (https://doi.org/10.1007/978-3-662-61184-5_7) contains supplementary material, which is available to authorized users.

The original version of this chapter was revised: The belated corrections have been incorporated. The correction to this chapter is available at https://doi.org/10.1007/978-3-662-61184-5_11

M. J. A. M. van Putten, *Dynamics of Neural Networks*,
https://doi.org/10.1007/978-3-662-61184-5_7

7.1.1 Background

The first recordings of EEG rhythms in human subjects were performed in the twenties of the previous century by the German psychiatrist Hans Berger. Since these early days, a wide variety of EEG rhythms has been discovered, ranging from alpha and beta rhythms during restful waking and gamma rhythms during cognitive and perceptual processing, to slow waves and alpha spindles during sleep. Also, most psychiatric and neurological syndromes have specific EEG signatures that can range from subtle alterations of normal rhythms to highly abnormal discharge patterns. Examples of such signatures are numerous and include increased gamma activity in schizophrenic patients, focal high-amplitude polymorphic delta waves in traumatic brain injury or space occupying lesions, and generalized spike-wave discharges during absence seizures.

Figure 7.1 shows three examples of EEG rhythms. In contrast to the cognitive, perceptual, and clinical correlates of EEG rhythms, which are well documented, the physiological mechanisms by which these rhythms are generated are incompletely understood at present. Advances in our understanding of the underlying physiological mechanisms are realized through an interplay between experiments and computational modeling. In this chapter we introduce a type of models called *neural mass models*. In contrast to the single-neuron models we have come across in previous chapters, neural mass models are *macroscopic* models in that they describe the *average behavior* of large populations of neurons. Since the EEG is also a macroscopic quantity, reflecting the average dendritic activity within local populations of cortical

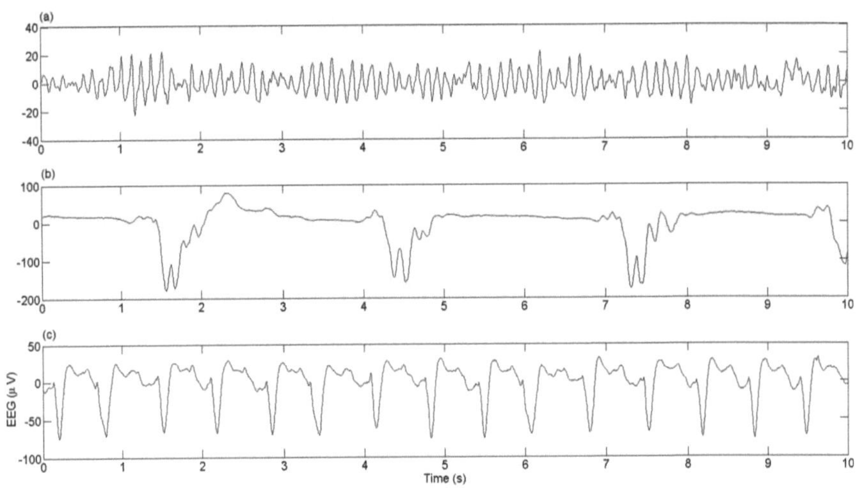

Fig. 7.1 Examples of EEG rhythms. **a** Spontaneous alpha oscillations recorded in a healthy subject during restful waking. **b** and **c** show a burst-suppression pattern and generalized period discharges, respectively, during postanoxic coma caused by oxygen shortage after cardiac arrest. Data from Medisch Spectrum Twente

pyramidal neurons, neural masses are naturally connected to the EEG. In the context of EEG rhythms, there are three main advantages of neural masses over network models of spiking neurons. First, a realistic network model of the EEG contains $\sim 10^5$ neurons which makes it unpractical to simulate on a computer. Second, the number of parameters in such network models is of the same order of magnitude as the number of neurons. Given that the values of most of these parameters are not precisely known, the search for parameters for which the model generates realistic EEG rhythms is tremendous. Third, due to their high dimensionality, network models do not easily allow underlying dynamical principles to be uncovered. Admittedly, this third drawback is only experienced as such by theoreticians and not so much by neurophysiologists. In contrast to network models, neural masses are low-dimensional, have few parameters, and allow for a dynamical analysis. One of the challenges in computational neuroscience is to understand the relationship between the parameters of models of spiking neural networks and neural mass models since this will shed light on how microscopic physiology shapes the large scale neural activation patterns that are observed in the EEG.

In this chapter we introduce neural mass modeling from a bottom-up viewpoint, where we start with the simplest case and add extra structure to build increasingly realistic models. However, we will restrict to models that can be analyzed analytically. In Sect. 7.2 we introduce the building blocks of every neural mass: the synaptic response and the activation function. In Sect. 7.2.3 we analyze the simplest neural mass model. In the analysis we treat steady states, linear stability, resonances, and the EEG power spectrum. In Sect. 7.3 we analyze the effect the effect of synaptic feedback. We will see that feedback can lead to bi-stable behavior. In Sect. 7.4 we consider a neural mass model that describes the behavior that results from synaptic interactions between excitatory and inhibitory neuronal populations. We will see that such reciprocal feedback leads to the emergence of rhythmic behavior. Throughout the text, we make use of Laplace transformations. The relevant definitions and properties can be found in Appendix A.

7.1.2 Connection with the EEG

A neural mass describes the behavior of a local population of neurons of a certain kind. This could be spiny stellate cells in cortical tissue, inhibitory neurons in thalamic reticular nuclei, pacemaker neurons in the brainstem, or any other kind of neurons. A neural mass model typically consists of several interconnected masses. To give an example, one of the neural mass models that is currently used to explain the existence of alpha oscillations comprises four interconnected neural masses modeling the behavior of cortical inhibitory and pyramidal neurons, thalamic reticular neurons, and thalamo-cortical relay neurons.

The state variable that describes the behavior of each of the masses within a neural mass model is the *mean membrane potential* $V(t)$ in mV of all neurons within the population. Although neural mass models of the EEG can be build up of several interconnected masses, the membrane potential of the mass consisting of cortical pyramidal neurons provides the link with the EEG. Remember that the EEG reflects a weighted sum of dendritic currents of cortical pyramidal neurons situated near the electrode. Thus, although all masses in the model contribute to the behavior of the mass consisting of cortical pyramidal neurons, it is only the behavior of the latter that is directly observable in the EEG. The mean membrane potential $V(t)$ of a neural mass consisting of cortical pyramidal neurons is approximately proportional to the EEG signal:

$$\mathrm{EEG}(t) \propto V(t). \tag{7.1}$$

In reality, the EEG signal is a low-pass filtered version of $V(t)$, caused by the propagation of $V(t)$ through cortical tissue, skull, and scalp. We will ignore this for the moment since it has no qualitative effect on the EEG and model the EEG signal by $V(t)$.

7.2 The Building Blocks

Any of the masses making up a neural mass model consists of two building blocks. The first building block is the *synaptic response* which describes how the mean firing rate $Q_{in}(t)$ in $1/\mathrm{s}$ of action potentials coming into the mass determines its mean membrane potential $V(t)$. The second building block is the *activation function* which describes how $V(t)$ determines the *mean firing rate* $Q(t)$ of the neural mass itself. Thus, a neural mass can be viewed as a system that converts pre-synaptic firing rates to post-synaptic firing rates. Figure 7.2 provides an illustration.

7.2.1 *The Synaptic Response*

The value of the mean membrane potential $V(t)$ of a neural mass is the result of all currents flowing through all dendrites of all neurons making up the neural mass. The net result of all these incoming currents is a filtering of $Q_{in}(t)$ whose properties are specified by the *synaptic response* $h(t)$, which is measured in $1/\mathrm{s}$. The synaptic response is a function that satisfies $h(t) = 0$ for $t < 0$ and is normalized to unity:

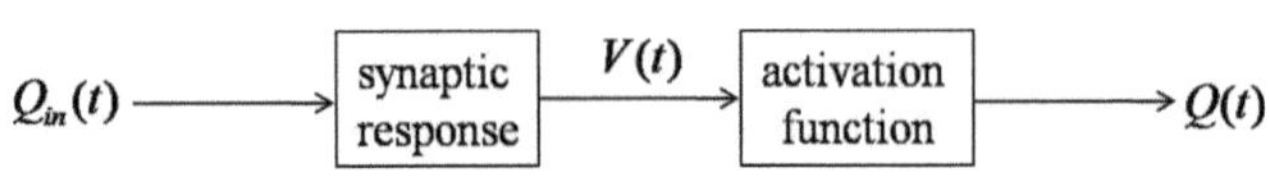

Fig. 7.2 A neural mass consists of two building blocks: the synaptic response and the activation function. The synaptic response describes how incoming firing rates determine the membrane potential of the mass and the activation function describes how the membrane potential determines its firing rate

$$\int_0^\infty h(t)dt = 1. \tag{7.2}$$

The mean membrane potential of the mass is given by the convolution of h with the incoming firing rate $Q_{in}(t)$:

$$V(t) = \nu h * Q_{in}(t) \stackrel{def}{=} \nu \int_{-\infty}^{t} h(t-\tau)Q_{in}(\tau)\,d\tau = \nu \int_0^\infty h(\tau)Q_{in}(t-\tau)\,d\tau, \tag{7.3}$$

where the *mean synaptic efficacy* of the synapses $\nu = Ns$ is defined as the product of the number of synaptic contacts N and the mean synaptic strength s measured in mVs. We need the history of the input to compute the potential at time t. As $\lim_{t\to\infty} = 0$, the influence of the history fades away as time elapses. Note that in the absence of input ($Q_{in}(t) = 0$) the mean membrane potential equals zero. In other words, we have set the mean resting potential of the mass to zero.

When $\nu > 0$, a positive incoming firing rate will depolarize the neural mass, in which case we refer to the synapses as *excitatory*. Similarly, when $\nu < 0$, the synapses are *inhibitory*. This is in contrast to previous chapters, where we have learned that a given synapse can be excitatory or inhibitory, depending on the reversal potentials of the receptors. Reversal potentials can be incorporated into the synaptic response in the following way:

$$V(t) = V_{\text{rest}} - \frac{V(t) - V_{\text{rev}}}{|V_{\text{rest}} - V_{\text{rev}}|}\nu h * Q_{in}(t), \tag{7.4}$$

where V_{rev} denotes the mean reversal potential in mV of all receptors within the population. When the range of $V(t)$ is small compared to $|V_{\text{rest}} - V_{\text{rev}}|$, the synaptic response can be approximated by (7.3). In the rest of this chapter we focus on the simple case given by (7.3). We note however, that several pharmacological agents directly affect the reversal potentials of specific receptors types. To model these effects, we need to explicitly incorporate a reversal potential as done in (7.4).

Electrophysiological experiments (see also Chaps. 1 and 2) show that a reasonable parametrization of h is the following[2]:

$$h(t) = \frac{\alpha\beta}{\beta - \alpha}\left[e^{-\alpha t} - e^{-\beta t}\right], \tag{7.5}$$

for $t \geq 0$, where $\alpha > 0$ and $\beta > 0$ are the *synaptic decay rate* and *synaptic rise rate* in 1/s, respectively. Typically, $\beta > \alpha$. You can check that h is indeed normalized to unity. Figure 7.3 provides an illustration. Commonly used limiting cases are discussed in the Exercises 7.2 and 7.3, and can be compared to the treatise in Chap. 2.

---

[2]Here, we use h for the synaptic response function. In Chaps. 1 and 2, we used the symbol g for the channel conductance.

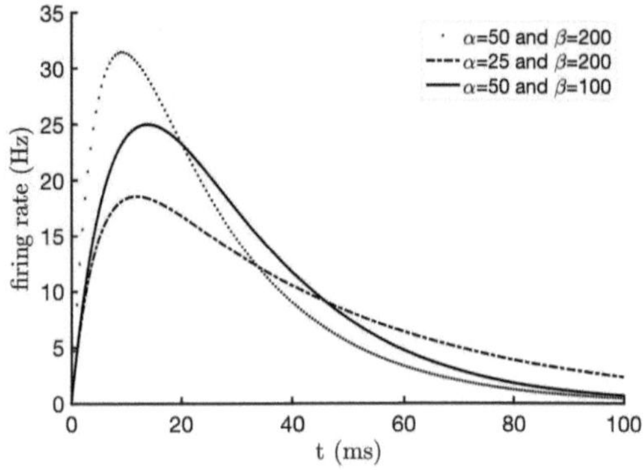

Fig. 7.3 Synaptic responses h as given by (7.5) for three different combinations of the synaptic decay and rise rates α and β, respectively. Note that varying the rate constants influences the maximal height of the synaptic response. This is due to the normalization of h

7.2.2 *The Activation Function*

The second building block of a neural mass is the *activation function* S. It describes how the mean membrane potential of the neural mass determines its mean firing rate. Formally, this relation is given by

$$Q(t) = S(V(t)). \tag{7.6}$$

Note that the activation function has no memory in the sense that the mean firing rate at time t does not depend on the membrane voltage at earlier times. Since the mean firing rate of a neural population increases as a function of its mean membrane potential, V, we will assume that

$$S'(V) > 0, \tag{7.7}$$

where $'$ denotes taking the derivative with respect to V. Note that we now dropped the explicit time dependence of the membrane potential. Also, due to the refractory periods, the mean firing rates cannot be arbitrarily high. Let's say they lie in the interval $(0, Q_{max})$ for certain $Q_{max} > 0$. Thus, we will assume that

$$\lim_{V \to -\infty} S(V) = 0, \tag{7.8}$$

and

$$\lim_{V \to \infty} S(V) = Q_{max}. \tag{7.9}$$

The most commonly used activation function is the *sigmoid*, which is defined as

$$S(V) = \frac{Q_{max}}{1 + e^{-(V-\theta))/\sigma}}, \tag{7.10}$$

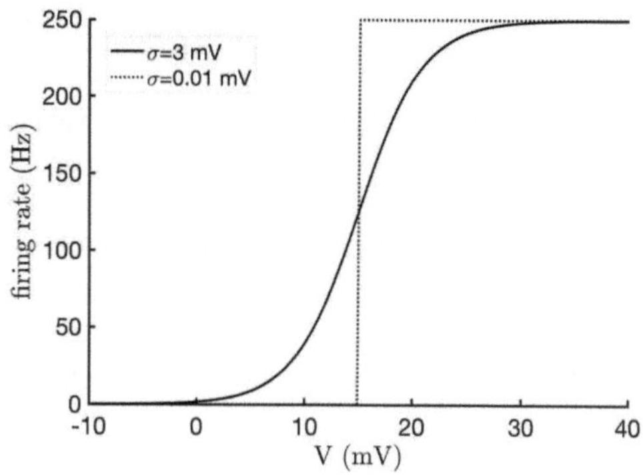

Fig. 7.4 Sigmoid activation function S for two different choices of σ ($\sigma = 3$ mV (solid line) and $\sigma = 0.01$ mV (dotted line). In both cases $\theta = 15$ mV and $Q_{max} = 250$ Hz. We will keep θ and Q_{max} fixed at these values throughout this chapter

where θ is the *mean firing threshold* in mV and σ models the dispersion of firing thresholds over the population. Note that when all neurons in the population have the same firing threshold, σ approaches 0 hence S reduces to the step function H_θ which is defined as

$$H_\theta(V) = \begin{cases} 0 & \text{if } V < \theta, \\ Q_{max} & \text{if } V \geq \theta. \end{cases} \tag{7.11}$$

Figure 7.4 illustrates a sigmoid activation function.

7.2.3 *Example*

Equations (7.3) and (7.6) together specify a neural mass. We now analyze the simplest case of a population of neurons that are not synaptically coupled to each other but only receive input from distant sources. In terms of the neural mass this means that $Q_{in}(t)$ does not depend on the mean membrane potential $V(t)$ but is generated by an independent process. We say that the neural mass has no *feedback*. Figure 7.5 provides an illustration. In Exercise 7.6 it is shown that if the synaptic response h has the form given by (7.5), then (7.3) can be re-written as the following second-order differential equation:

$$\ddot{V}(t) + (\alpha + \beta)\dot{V}(t) + \alpha\beta V(t) = \alpha\beta\nu Q_{in}(t), \tag{7.12}$$

where $\dot{V}$ denotes differentiation of V to time t. When $Q_{in}(t)$ is constant ($Q_{in}(t) = q$), we can deduce from (7.12) that the steady state V^* of $V(t)$ is given by $V^* = \nu q$. Notice that for very large positive values of q the mean firing rate of the neural mass approaches its maximum Q_{max}. Since the EEG measures potential *differences*, the steady state of $V(t)$ cannot be measured. However, neural populations *in vivo* are continuously subject to spike trains coming in from other neuronal structures. Assuming that the incoming spike trains are large in number and uncorrelated, $Q(t)$ can be approximated by a stochastic process. Thus, we write $Q_{in}(t) = q + Q_0(t)$

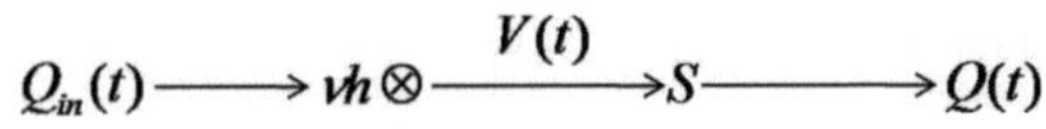

Fig. 7.5 Schematic diagram of a neural mass without feedback

where $Q_0(t)$ is a stochastic processes with mean zero. This means that, for every t, $Q_0(t)$ is a random variable with expectation zero. The equation for the fluctuations $\bar{V}(t) = V(t) - V^*$ of $V(t)$ about its steady state V^* can be derived from (7.3). Subtracting the identities $V(t) = \nu h * Q_{in}(t)$ and $V^* = \nu h * q$ yields and is given by

$$\bar{V}(t) = \nu h * Q_0(t). \tag{7.13}$$

The stability of the fluctuations in V is determined by the *resonances* λ, which are defined as the poles of the transfer function H from Q_0 to $\bar{V}$, which is defined as

$$H(s) = \frac{\tilde{\bar{V}}(s)}{\tilde{Q}_0(s)}, \tag{7.14}$$

is given by

$$H(s) = \frac{\nu\alpha\beta}{(s+\alpha)(s+\beta)}. \tag{7.15}$$

Since the resonances equal $\lambda = -\alpha, -\beta$ have negative real part, the fluctuations of V about V^* are stable. However, since the resonances have no imaginary part, the fluctuations $\bar{V}$ do not display periodic behavior.

When we assume that $Q_0 = \delta\xi(t)$, where $\xi(t)$ is a zero-mean and unit-variance white-noise process and $\delta > 0$, the EEG power spectrum

$$P_{\mathrm{EEG}}(\omega) = |H(i\omega)Q_0(i\omega)|^2, \tag{7.16}$$

is given by

$$P_{\mathrm{EEG}}(\omega) = \frac{(\alpha\beta\nu\delta)^2}{(\alpha^2+\omega^2)(\beta^2+\omega^2)}, \tag{7.17}$$

where we have used that $\tilde{\xi}(s) = 1$, where $\tilde{\xi}$ denotes the Laplace transformation of ξ. This means that the fluctuations contain all frequencies in an equal amount. For $\omega = 0$ the power spectrum reduces to $(\nu\delta)^2$, which equals the variance of the fluctuations in $\bar{V}$. Notice that P_{EEG} decreases as a function of ω. This means that the neural mass acts as a low-pass filter on the incoming fluctuations $Q_0(t)$. Moreover, the filter characteristics are completely determined by the synaptic rate constants α and β. In particular, they are *independent* of q. Figures 7.6 and 7.7 provide illustrations. Note the absence of a peak in the EEG power spectrum, reflecting the fact that this mass cannot produce rhythmic behavior. Remember that this neural mass models a population of neurons without internal synaptic contacts. Massive diffuse synaptic

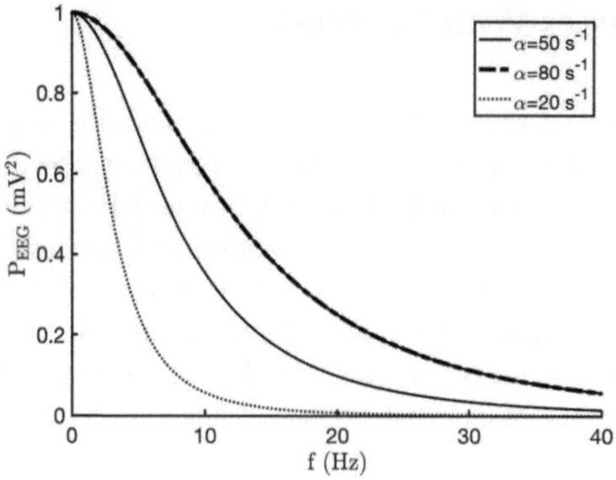

Fig. 7.6 EEG power spectrum of the neural mass given by (7.3) for different values of the synaptic decay rate α ($\alpha = 50\ s^{-1}$ (solid line), $\alpha = 80\ s^{-1}$ (dashed line), and $\alpha = 20\ s^{-1}$ (dotted line). In all three cases, the synaptic rise rate β is set to $\beta = 4\alpha$ and $\nu = 1$ mVs, $\delta = 1\ s^{-1}$. Note that $f(0) = (\nu\delta)^2 = 1$ mV^2

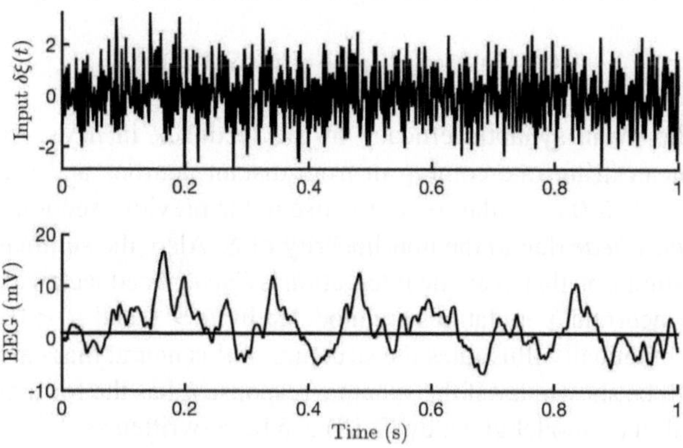

Fig. 7.7 Top: numerical simulation of the input $\delta\xi(t)$ to the neural mass and (bottom) the resulting EEG signal. $\beta = 200\ \mathrm{s}^{-1}$, $\alpha = 50\ \mathrm{s}^{-1}$, $\nu = 1$ mVs, $\delta = 1\ \mathrm{s}^{-1}$ and $q = 1\ s^{-1}$. Notice that due to the low-pass filtering properties of the mass, the high frequency fluctuations in the input are attenuated, hence are not visible in the EEG; the steady state value $V^* = \nu q = 1$mV, indicated by the horizontal line in the lower panel

damage as may be observed in patients with severe metabolic encephalopathies, for instance after cardiac arrest, may serve as an (almost) limiting clinical case in which we indeed can observe EEG spectra devoid of spectral peaks. This is a good moment to make Exercises 7.7–7.9.

7.3 Neural Masses With Feedback

In the Sect. 7.2 we have discussed how the average behavior of a large population of neurons can be modeled by a neural mass. This neural mass, however, did not take into account synaptic connections within the population which are present in real cortical tissue. In this section we describe how such *neuronal feedback* can be incorporated into the simple neural mass. Our analysis will further show that feedback within a population of excitatory neurons, but not inhibitory neurons, can lead to *bistability*, in which case the neuron mass can make sudden transitions between stable steady states under the influence of external stochastic fluctuations.

7.3.1 *Model Equations*

Neuronal feedback is modeled by adding a feedback term to the right-hand-side of (7.3):

$$V(t) = \nu h * Q_{in}(t) + \mu h * S(V(\cdot))(t), \tag{7.18}$$

where μ is the mean synaptic efficacy of the feedback in mVs. The input Q_{in} denotes the mean firing rate coming in from distant neurons and we will assume that $Q_{in}(t) = q + \delta\xi(t)$, similar to our treatise in the previous section. Note that the feedback is *non-linear* due to the non-linearity of S. Also, the summation in (7.18) reflects the assumption that synaptic integration is *linear*. Feedback within excitatory and inhibitory neuronal populations are modeled by $\mu > 0$ and $\mu < 0$, respectively. Figure 7.8 schematically illustrates the structure of this neural mass model. In Exercise 7.10 it will be shown that if the synaptic response h has the form given by (7.5), then the neural mass model given by (7.18) can be re-written as

$$\ddot{V}(t) + (\alpha + \beta)\dot{V}(t) + \alpha\beta V(t) = \alpha\beta\nu Q_{in}(t) + \alpha\beta\mu S(V(t)). \tag{7.19}$$

7.3.2 *Steady States*

From (7.19) we deduce that for constant firing rate $Q_{in}(t) = q$, the steady states V^* of V satisfy

$$V^* = \nu q + \mu S(V^*), \tag{7.20}$$

hence correspond to the intersection points of the activation function $S(V)$ and the straight line $l(V) = -\nu q/\mu + V/\mu$. This is illustrated in Fig. 7.9.

From Exercise 7.12 and Figs. 7.8 and 7.9 we can deduce that if $\mu > 4\sigma/Q_{max}$ there exists an intermediate range of values of q for which V has *three possible steady-states*. In fact, only the lowest and highest steady-states are stable. This can also be

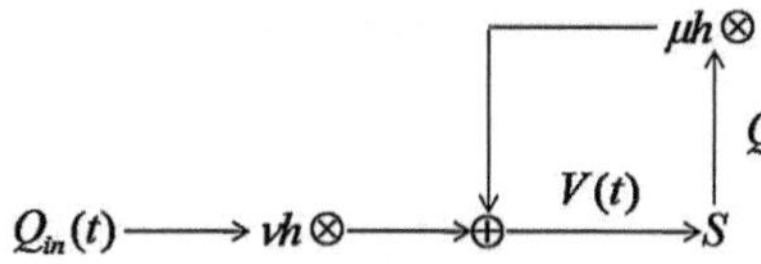

Fig. 7.8 Schematic diagram of a neural mass with feedback

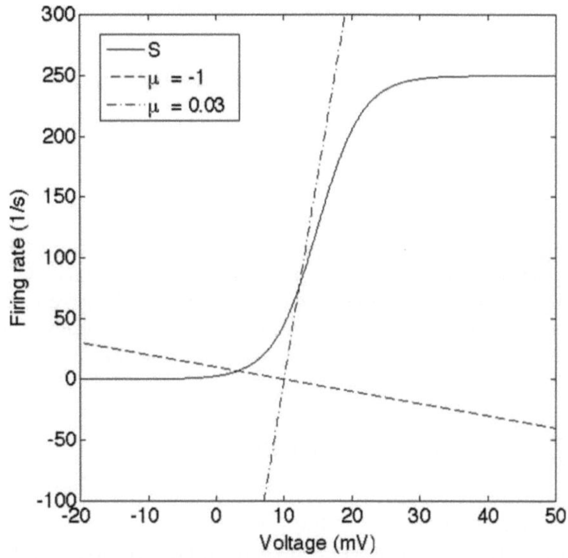

Fig. 7.9 Shown are the activation function S, together with the line $l(V)$ for two choices of the feedback efficacy μ, namely $\mu = -1$ (dashed line) and $\mu = 0.03$ (dash-dotted line). In the first case the feedback is inhibitory and in the second case the feedback is excitatory. In both cases we chose $q = 10\ (1/s)$ and $\nu = 1$ mVs

argued as follows: if the highest steady state were unstable for fixed input $Q(t) = q$ and μ, $V(t) \to \infty$, when initialized appropriately. As a consequence, $S(V(t)$ would approach Q_{max}, hence according to (7.18) $V(t)$ would approach $\nu q + \mu Q_{max}$, which contradicts the initial assumption. In the same way you can show that the lowest steady state is stable. Consequently, the middle steady state is unstable. Thus, when the neural mass has three steady states, it is *bi-stable*.

When the input to the neural mass is constant ($Q_{in}(t) = q$) its membrane voltage will approach one of the stable steady states and remain there. However, due to stochastic fluctuations Q_0 of Q_{in}, which are always present in real neural populations, $V(t)$ might suddenly switch between the stable steady states. Figure 7.10 illustrates that stochastic fluctuations in $Q_{in}(t)$ can cause $V(t)$ to jump from the lower to the upper stable steady state, when initialized close to the lower steady state. Moreover, it illustrates that such jumps are more probable and occur earlier for higher noise intensities δ. Such *bi-stable behavior* can be observed in cortical pyramidal neurons, for example during deep sleep stages, general anesthesia, and coma. The high firing rates are however, physiologically unsustainable and are suppressed by intrinsic neuronal mechanisms that are not incorporated into our neural mass.

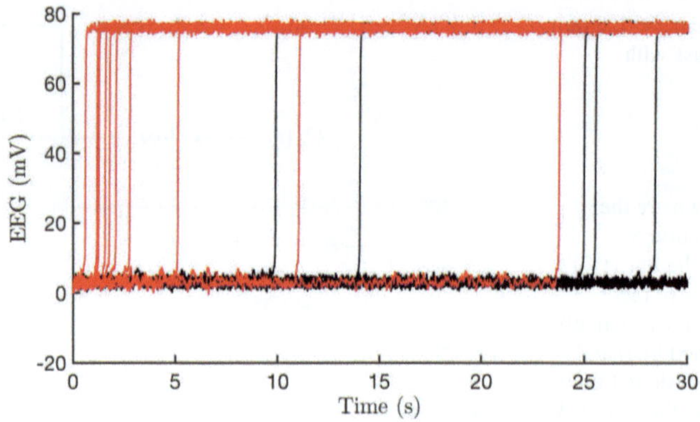

Fig. 7.10 Shown are 10 repeats of simulated time-series of the potential $V(t)$ for $\delta = 0.12$ mV (black lines) and for $\delta = 0.15$ mV (red lines). In the simulations we chose $\nu = 1$ mVs, $\mu = 0.3$ mVs and $q = 1$ Hz. Moreover, $V(t)$ was initialized to the lowest stable steady state. The vertical lines are the switches from the lower to the higher stable steady state. Note that for the larger value of the intensity of the stochastic fluctuations, δ, the switches occur more often and earlier

7.3.3 Linear Approximation

When the intensity δ of the stochastic fluctuations in $Q_{in}(t)$ is small, switches of $V(t)$ to a different steady state are rare and the voltage will fluctuate about one steady state for most of the time. In addition, if the noise intensity is small as compared to the variance of the spike threshold over the population, the activation function S can be approximated by a linear function. Thus, for small fluctuations of V about V^* we approximate S using a first-order Taylor expansion around V^*:

$$S(V) \approx S(V^*) + \frac{d}{dV} S(V^*)(V - V^*). \tag{7.21}$$

Substituting (7.21) in (7.18) gives the linearized dynamics of V:

$$V(t) = h * \left[\nu Q_{in}(t) + \mu S(V^*) + G(V(t) - V^*)\right], \tag{7.22}$$

where we have defined the *feedback gain*

$$G = \mu \frac{d}{dV} S(V^*), \tag{7.23}$$

which is measured in mV/s. Using the solution of Exercise 7.5 we can also write for the feedback gain, G

$$G = \mu \frac{d}{dV} S(V^*) = \frac{\mu}{\sigma} S(V)(1 - S(V)/Q_{\max}). \tag{7.24}$$

The feedback gain is proportional to both the efficacy of the synaptic feedback μ and the *excitability* $dS/dV(V^*)$ of the mass at its steady state voltage V^*. Note that the neural mass is most excitable when its steady state voltage equals its mean spike threshold ($V^* = \theta$). The dynamics of the fluctuations $\bar{V}(t) = V(t) - V^*$ about V^* are described by

$$V(t) = h * \big[\nu Q_{in}(t) + \mu S(V^*) + G(V(t) - V^*)\big], \tag{7.25}$$

$$\bar{V}(t) = h * \big[\nu Q_{in}(t) + \mu S(V^*) + G(V(t) - V^*)\big] \overbrace{-\nu q - \mu S(V^*)}^{-V^*}, \tag{7.26}$$

$$= h * \big[\nu Q_0(t) + G\bar{V}(t)\big], \tag{7.27}$$

where we have made use of (7.20), and the fact that the convolution of h with a constant is equal to that constant since $\int_0^\infty h(\tau)\,d\tau = 1$.

7.3.4 Resonances

To determine the stability of the fluctuations in V, we first calculate the transfer function H from Q_0 to $\bar{V}$. It is derived by Laplace transforming (7.27), which gives

$$\tilde{\bar{V}}(s) = h(s)\big[\nu Q_0(s) + G\tilde{\bar{V}}(s)\big], \tag{7.28}$$

and can be rearranged to obtain

$$H(s) == \frac{\nu\alpha\beta}{(s+\alpha)(s+\beta) - \alpha\beta G}. \tag{7.29}$$

It is now possible, using (7.29), to calculate resonances. This is left as Exercise 7.15. For excitatory gain ($G > 0$) the resonances have no imaginary part and when G reaches 1, they reduce to 0 and $-(\alpha + \beta)$. Since the first resonance has real part zero, 1 is the value of G for which the neural mass *destabilizes*. When this happens, $V(t)$ makes a sudden switch between its two stable steady states, discussed further in Exercises 7.16 and 7.17.

7.3.5 EEG Power Spectrum

When the resonances are stable, we can compute the EEG power spectrum that results from white-noise fluctuations $\delta\xi(t)$ in Q_0:

$$P_{\text{EEG}}(\omega) = \left| \frac{\alpha\beta\nu\delta}{(i\omega + \alpha)(i\omega + \beta) - \alpha\beta G} \right|^2. \tag{7.30}$$

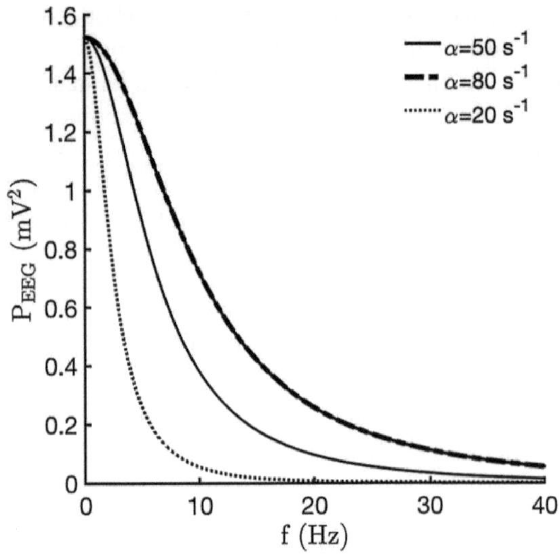

Fig. 7.11 EEG power spectrum of the neural mass with feedback for different values of the decay rate α. In all three cases, $\beta = 4\alpha$. The efficacy of the feedback, $\mu = 0.2$ mVs, $\nu = 1$ mVs, $q = 0$ Hz and $\delta = 1$ Hz. Spectral peak frequencies, indicating oscillations, are absent. Note also, that an enhancement of the frequencies occurs at low frequencies; in Exercise 7.19 this is further explored

The power spectrum for three different values of α is shown in Fig. 7.11. Further characteristics are explored in Problems 7.18 and 7.19.

7.4 Coupled Neural Masses

Any realistic model of the generation of EEG rhythms must include at least two kinds of neuronal populations: excitatory and inhibitory ones. Although in real cortical tissue there can be found several kinds of inhibitory neurons with different electrophysiological properties, we restrict to modeling one kind. We will see that coupled inhibitory and excitatory neural masses give rise to oscillations in the EEG.

7.4.1 Model Equations

We consider a neural mass consisting of cortical pyramidal neurons, which is synaptically coupled to a neural mass consisting of inhibitory neurons. Their mean membrane potentials are denoted by $V_e(t)$ and $V_i(t)$, respectively. This neural mass is specified by the following equations:

$$V_e(t) = h * \nu Q_{in}(t) + h * \mu_{ie} S(V_i(\cdot))(t), \tag{7.31}$$

$$V_i(t) = h * \mu_{ei} S(V_e(\cdot))(t), \tag{7.32}$$

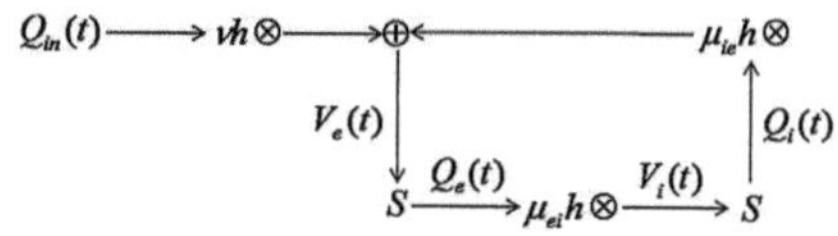

Fig. 7.12 Schematic diagram of two coupled neural masses

where $V_e(t)$ and $V_i(t)$ are the mean membrane potentials of the excitatory and inhibitory masses, respectively, $Q(t) = q + Q_0(t)$ as in previous sections, and $\mu_{ie} < 0$ and $\mu_{ei} > 0$ are the mean efficacies of the inhibitory and excitatory-to-pyramidal synapses, respectively. Note that only the mass consisting of pyramidal neurons receives non-specific input $Q_{in}(t)$. Figure 7.12 schematically illustrates the structure of this neural mass model. In Exercise 7.20 you should show that this neural mass model can be rewritten as a system of two second-order differential equations.

7.4.2 Steady-States

For constant input $Q_{in}(t) = q$, the steady states (V_e^*, V_i^*) of (V_e, V_i) satisfy the coupled equations

$$V_e^* = \nu q + \mu_{ie} S(V_i^*), \tag{7.33}$$
$$V_i^* = \mu_{ei} S(V_e^*), \tag{7.34}$$

which can be combined to yield a closed equation for the steady states of V_e:

$$0 = -V_e^* + \nu q + \mu_{ie} S(\mu_{ie} S(V_e^*)). \tag{7.35}$$

Note that the steady states of V_e correspond to the intersection points of the straight line $V_e \to -\nu q/\mu_{ie} + V_e/\mu_{ie}$ and the function $V_e \to S(\mu_{ie} S(V_e))$.

7.4.3 Linear Approximation

For small fluctuations of V_e and V_i about their steady states V_e^* and V_i^*, (7.31) and (7.32) can be linearized about V_i^* and V_e^*, respectively, yielding

$$V_e(t) = h * \left[\nu Q(t) + \mu_{ie} S(V_i^*) + G_{ie}(V_i(t) - V_i^*)\right],$$
$$V_i(t) = h * \left[\mu_{ei} S(V_e^*) + G_{ei}(V_e(t) - V_e^*)\right],$$

where we have defined the *gains* G_{ie} and G_{ei} by

$$G_{ie} = \mu_{ie}\frac{d}{dV}S(V_i^*), \tag{7.36}$$

and

$$G_{ei} = \mu_{ei}\frac{d}{dV}S(V_e^*). \tag{7.37}$$

The linearized dynamics can be re-written in terms of the fluctuations $\bar{V}_e(t) = V_e(t) - V_e^*$ and $\bar{V}_i(t) = V_i(t) - V_i^*$:

$$\bar{V}_e(t) = h * \left[\nu Q_0(t) + G_{ie}\bar{V}_i(t)\right], \tag{7.38}$$

$$\bar{V}_i(t) = h * G_{ei}\bar{V}_e(t), \tag{7.39}$$

where we have made use of (7.33) and (7.34).

7.4.4 Resonances

To calculate the resonances we first compute the transfer function H from Q_0 to $\bar{V}_e$. It is given by

$$H(s) = \frac{\nu\tilde{h}(s)}{1 - G_T\tilde{h}^2(s)}, \tag{7.40}$$

where $\tilde{h}(s)$ denotes the Laplace transformation of h and where $G_{eie} = G_{ei}G_{ie}$ denotes the gain in the feedback loop. Note that $G_{eie} \leq 0$. To calculate the resonances λ, we note that they satisfy

$$1 - G_{eie}\tilde{h}^2(\lambda) = 0, \tag{7.41}$$

which is equivalent to

$$\left[(s+\alpha)(s+\beta)\right]^2 = G_{eie}(\alpha\beta)^2, \tag{7.42}$$

hence can be re-written as

$$(s+\alpha)(s+\beta) = \pm i\alpha\beta\sqrt{|G_{eie}|}. \tag{7.43}$$

Thus, the resonances are given by

$$\lambda = -\frac{\alpha+\beta}{2} \pm \sqrt{\left(\frac{\alpha-\beta}{2}\right)^2 \pm i\alpha\beta\sqrt{|G_{eie}|}}. \tag{7.44}$$

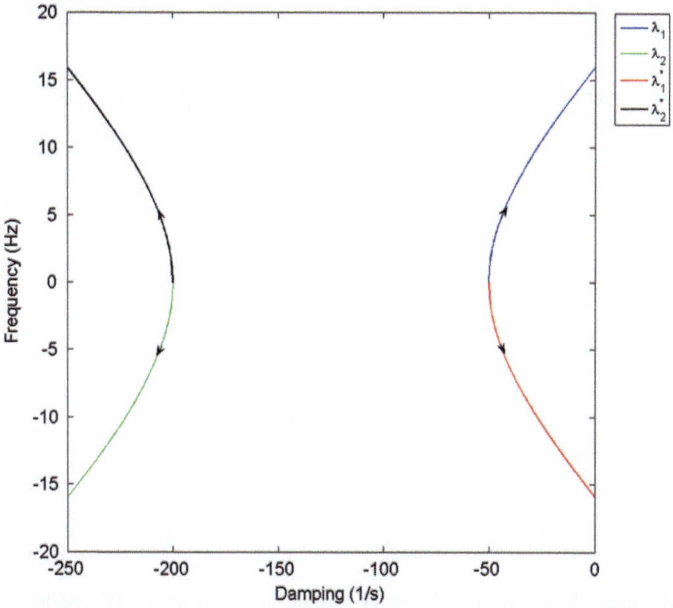

Fig. 7.13 Trajectories of the resonances in the complex plane when $|G_{eie}|$ is increased from zero to its critical value $G_{crit} = (\alpha + \beta)^2/\alpha\beta$. We chose $\alpha = 50$ 1/s and $\beta = 200$ 1/s. Note that the poles come in complex conjugate pairs (indicated by * in the figure's legend) and that they equal $-\alpha$ and $-\beta$ for $G_{eie} = 0$

Assuming the s is purely imaginary in (7.43) gives the solution $s = \sqrt{\alpha\beta}$ for $G_{eie} = (\alpha + \beta)^2/\alpha\beta$. Since for this *critical value* $G_{crit} = (\alpha + \beta)^2/\alpha\beta$ of G_{eie} one pair of complex conjugate resonances crosses the imaginary axis, its steady state becomes unstable and makes way for a limit cycle via a Hopf bifurcation, as discussed in Chap. 4. Moreover, the frequency of the resulting oscillations equals $\sqrt{\alpha\beta}$. For typical values $\alpha = 50\ \text{s}^{-1}$ and $\beta = 200\ \text{s}^{-1}$ the oscillations have a frequency of $\sqrt{\alpha\beta}/2\pi \approx$ 16 Hz. Figure 7.13 shows the trajectories of the resonances in the complex plane when G_{eie} is increased from zero to G_{crit}.

7.4.5 EEG Power Spectrum

Assuming that $Q_0 = \sigma\xi(t)$, the EEG power spectrum is given by

$$P_{EEG}(\omega) = \left| \frac{\sigma\nu\tilde{h}(i\omega)}{1 - G_{eie}\tilde{h}^2(i\omega)} \right|^2. \tag{7.45}$$

However, (7.45) is only defined when the fluctuations of V_e about its resting state are stable. This is equivalent to all resonances having a negative real part. Figure 7.14

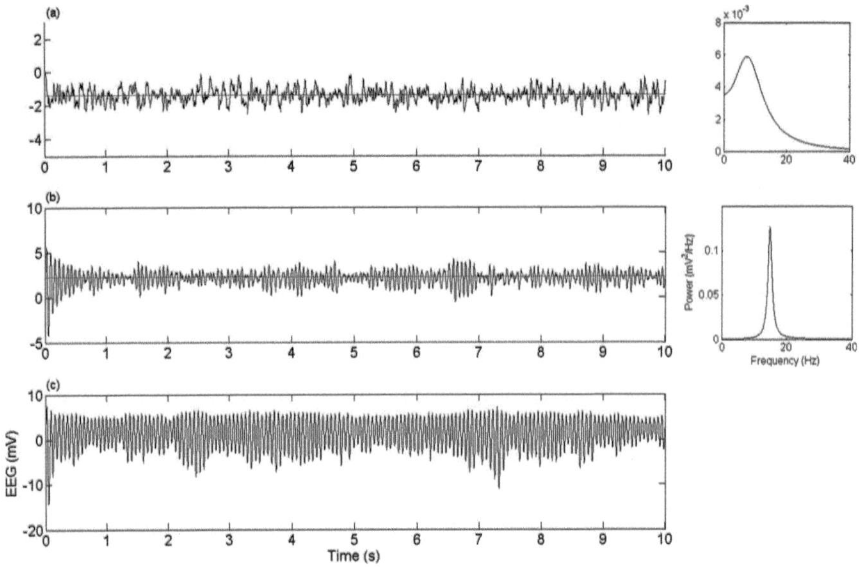

Fig. 7.14 Shown are three simulated EEG time-series using (7.50) and (7.51) and their corresponding power spectra. **a** $q = 3$, **b** $q = 14$, and **c** $q = 17$. The horizontal red lines denote the steady state value of the EEG. In (**a**) and (**b**) the feedback gain G_{eie} is below its critical value G_{crit}, while in (**c**) it exceeds its critical value. Other parameter values were constant in all three cases and set as follows: $\alpha = 50$ 1/s, $\beta = 200$ 1/s, $\nu = 1$, $\mu_{ei} = -\mu_{ie} = 1$, and $\delta = 0.1$

shows simulated EEG time-series with their corresponding power spectra for three increasing values of q. In all three simulation the mean efficacies of the synapses were set to $\mu_{ei} = -\mu_{ie} = 1$. Figure 7.14a illustrates that synaptic coupling gives rise to waxing and waning oscillations of about 8 Hz resembling alpha oscillations observed in EEG recordings of healthy subjects during rest (see Fig. 7.14a). The oscillations increase in amplitude and frequency with increasing q as illustrated in Fig. 7.14b. Finally, when the feedback gain G exceeds its critical value G_{crit} the mass destabilizes and gives rise to *pathological oscillations*. In the context of neural mass modeling, the boundary between physiological EEG rhythms and pathological discharges such as those seen during coma and epileptic seizures often corresponds to the boundary between small (linear) fluctuations about stable steady states and (non-linear) limit-cycle oscillations.

7.5 Modeling Pathology

Meanfield models have been applied to further our understanding of pathology, as well. This includes models that simulate EEG patterns in seizures and hypoxia/ ischaemia, to be discussed in Chaps. 9 and 8.

7.6 Summary

We discussed neural mass (or meanfield) models in relation to the EEG. Neural mass models describe electrical potentials as collectively generated by many neurons. These models assist in understanding how collective behaviour of neurons generate electrical rhythms, as recorded with the EEG. Further, these models can propose candidate mechanisms involved in various pathologies, including seizures and hypoxic ischaemic brain damage, discussed in later chapters.

Problems

7.1 Show that the right-hand-side of (7.3) has the unit of mV.

7.2 Consider (7.5). Show that $\lim_{\beta\to\alpha} h(t) = \alpha^2 t e^{-\alpha t}$ (hint: write $\beta = \alpha + \epsilon$ for certain $\epsilon > 0$ and apply l'Hôpital's rule). In this limit, $h(t)$ is called the *alpha function*.

7.3 Show that if β in (7.5) is much larger than α ($\beta \gg \alpha$) $h(t)$ can be approximated as $h(t) = \alpha e^{-\alpha t}$.

7.4 The anesthetic agent *propofol* acts by binding to $GABA_A$ receptors. Its effect is a decrease in the rate of receptor de-activation, leading to longer lasting inflow of Cl^- and thereby to a prolonged post-synaptic potential. The maximal current flow however, remains unchanged. How would you incorporate the action of propofol in the synaptic response h? (hint: define an appropriate normalization constant for h).

7.5 Consider (7.10). Show that $S'(V) = S(V)\left(1 - S(V)/Q_{max}\right)/\sigma$, where $'$ denotes the derivative with respect to V.

7.6 Show that if the synaptic response h has the form given by (7.5), then (7.3) can be re-written as the following second-order differential equation:

$$\ddot{V}(t) + (\alpha + \beta)\dot{V}(t) + \alpha\beta V(t) = \alpha\beta\nu Q_{in}(t), \tag{7.46}$$

where $\dot{V}$ denotes differentiation of V to time t.

7.7 In this exercise we compute the EEG power spectrum of the neural mass without feedback, exemplified in Sect. 7.2.3, in a different way. Suppose the neural mass is driven by an oscillatory signal with amplitude δ and angular frequency ω_0: $Q_{in}(t) = \delta e^{i\omega_0 t}$. To compute the EEG power spectrum, we assume that the resulting membrane potential $V(t)$ has the following form: $V(t) = Ae^{i\omega_0 t}$ for certain complex-valued amplitude. In general, A depends on the frequency ω_0 of the driving signal. Use (7.46) to show that the power $|A|^2$ of $V(t)$ (viewed as a function of ω_0) equals $P_{EEG}(\omega_0)$.

7.8 A large group of *peptide neurotoxins* acts on the nervous system by blocking excitatory (glutamate) receptors. These include α-agatoxins form the funnel web spider, NSTX-3 from the orb weaver spider, and β-philanthotoxin from wasp venom. Suppose we record the EEG from a neural population that can be described by the neural mass discussed in Sect. 7.2.3. What happens to the EEG signal if we gradually perfuse the population with one of the above peptide neurotoxins? Illustrate this in a numerical simulation (use the Matlab file mass1.m).

7.9 Suppose we measure the EEG from the neural mass without feedback. Is it possible to determine—on the basis of the EEG power spectrum—to estimate δ? And what about α and β? And if we know, for example through invasive recordings, that $\beta = 4\alpha$?

7.10 Show that if the synaptic response h has the form given by (7.5), then the neural mass model given by (7.18) can be re-written as the following second-order differential equation:

$$\ddot{V}(t) + (\alpha + \beta)\dot{V}(t) + \alpha\beta V(t) = \alpha\beta\nu Q_{in}(t) + \alpha\beta\mu S(V(t)). \tag{7.47}$$

7.11 Consider (7.20) and Sect. 7.3 discussing neural masses with feedback. Argue that if the feedback is inhibitory ($\mu < 0$), the neural mass has exactly one steady-state for every q.

7.12 Consider (7.20) and Sect. 7.3 discussing neural masses with feedback.

a Argue that the neural mass has exactly one steady-state for every value of q if and only if $dl/dV > dS/dV(\theta)$.
b Show that this condition is equivalent to $\mu < 4\sigma/Q_{max}$.
c Recall that the second order differential equation (7.20) can also be written as a system of 2 first order equations by setting $V = x_1$ and $\dot{V} = x_2$. Show that this results in

$$\dot{x_1} = x_2$$
$$\dot{x_2} = -(\alpha + \beta)x_2 - \alpha\beta x_1 + \alpha\beta\nu q + \alpha\beta\mu S(V).$$

For the equilibria, it holds that $\dot{x_1} = \dot{x_2} = 0$. This will also result in (7.20).
d Show graphically that three equilibria exist for the condition $\mu > 4\sigma/Q_{\max}$, assuming that q is sufficiently large.

7.13 Consider (7.20) and Sect. 7.3 discussing neural masses with feedback. Argue that for very high firing rates q, the neural mass has a maximal firing rate and that for very low (negative) values of q that the neural mass is silent.

7.14 Oxygen shortage in nervous tissue, for example as might occur after cardiac arrest, leads to a dysfunction of the Na^+ and K^+ pumps of the neurons. Consequently, the resulting influx of Na^+ and outflux of K^+ increases their resting potential. We can incorporate this hypoxic effect into the neural mass model by decreasing the mean spike threshold θ. What is the effect of hypoxia on the mean firing rate of the neural mass? (use the Matlab file mass2.m with the following parameter settings: $\mu = 0.1$, $\nu = 1$, $q = 0$, and $\delta = 0.1$).

7.15 Show that the resonances are given by

$$\lambda = -\frac{\alpha + \beta}{2} \pm \frac{1}{2}\sqrt{(\alpha - \beta)^2 + 4\alpha\beta G}.$$

7.16 Show that the neural mass destabilizes exactly when the steady-state firing rate $Q^* = S(V^*)$ reaches one of the following values:

$$Q^* = \frac{Q_{max}}{2}\left[1 \pm \sqrt{1 - \frac{4\sigma}{\mu Q_{max}}}\right]. \tag{7.48}$$

7.17 Show that when we increase inhibitory gain ($G < 0$) the two resonances approach each other and collide at $-(\alpha + \beta)/2$ when G reaches $-(\alpha - \beta)^2/4\alpha\beta$.

7.18 Show that in the absence of feedback ($G = 0$) (7.30) reduces to (7.17).

7.19 Show that the EEG power spectrum can be re-written as

$$P_{\text{EEG}}(\omega) = \frac{(\nu\alpha\beta\delta)^2}{(\alpha^2 + \omega^2)(\beta^2 + \omega^2) + 2(\omega^2 - \alpha\beta)G + G^2}. \tag{7.49}$$

Note that the frequency-dependent effect of the feedback is an attenuation of EEG power for $\omega > \sqrt{\alpha\beta}$ and an enhancement for $\omega < \sqrt{\alpha\beta}$. Given that $\alpha = 50$ 1/s and $\beta = 200$ 1/s are typical values observed in neural tissue, the cut-off frequency is about $\omega/2\pi = \sqrt{\alpha\beta}/2\pi \approx 16$ Hz.

7.20 Show that the neural mass model expressed in (7.31) and (7.32) can be re-written as the following set of coupled second-order differential equations:

$$\ddot{V}_e(t) + (\alpha + \beta)\dot{V}_e(r) + \alpha\beta V_e(t) = \alpha\beta\nu Q(t) + \alpha\beta\mu_{ie} S(V_i(t)), \tag{7.50}$$

$$\ddot{V}_i(t) + (\alpha + \beta)\dot{V}_i(r) + \alpha\beta V_i(t) = \alpha\beta\mu_{ei} S(V_e(t)). \tag{7.51}$$

7.21 Argue that for all values of q, V_e has exactly one steady state in (7.35).

7.22 Show that when either the excitatory or inhibitory synapses are blocked, the EEG power spectrum reduces to (7.17).

7.23 The psychoactive drug *diazepam* has anxiolytic, amnesic, and anticonvulsant properties. It acts by increasing the maximal current flow through GABA_A receptors. How would you incorporate its action in the neural mass model described in Sect. 7.4?

7.24 What is the effect of benzodiazepines as diazepam on the spontaneous alpha rhythm? (use the Matlab files mass3.m and mass3-spectrum.m). (suppose that the baseline parameters are given by $\alpha = 50\ 1/\text{s}$, $\beta = 200\ 1/\text{s}$, $\nu = 1$, $\mu_{ei} = -\mu_{ie} = 1$, $\delta = 0.1$, and $q = 7\ 1/\text{s}$).

7.25 Is a benzodiazepine able to abort the pathological limit cycle dynamics observed in Fig. 7.14?

Part V
Pathology

Chapter 8
Hypoxia and Neuronal Function

In all serious disease states we find a concomitant low oxygen state... Low oxygen in the body tissues is a sure indicator for disease...Hypoxia, or lack of oxygen in the tissues, is the fundamental cause for all degenerative disease. Oxygen is the source of life to all cells.

— Steven Levine

8.1 Introduction

The brain is obligatory dependent on sufficient oxygen and glucose.[1] Deprivation of either glucose or oxygen (oxygen-glucose deprivation, OGD) will result in abnormal functioning of neurons, and if sufficiently severe, will ultimately result in irreversible damage. The extent and nature of the neuronal damage is varied, and ranges from subtle changes in synaptic function to neuronal cell death. Under normal conditions, the cerebral blood flow (CBF) is typically 750 milliliters per minute or 15% of the cardiac output. This equates to 50–54 ml of blood per 100 grams of brain tissue per minute [91].

The cascade of events that takes place during OGD can be globally described as follows: initially, synaptic function fails, since this is one of the most energy

[1]That air is essential for living was discovered about 350 years ago by Robert Boyle. In his experiments (assisted by Robert Hook) he proved that animals die if deprived from air. A few years later, John Mayow discovered that it was only a fraction of the air that is actually used for breathing or combustion. In 1772–1774, Joseph Priestley and Carl Scheele independently isolated oxygen from air, but it was Lavoisier who proved the relevance of the "new gas" and called it oxygen. These historical details were found in [104].

Electronic supplementary material The online version of this chapter (https://doi.org/10.1007/978-3-662-61184-5_8) contains supplementary material, which is available to authorized users.

The original version of this chapter was revised: The belated corrections have been incorporated. The correction to this chapter is available at https://doi.org/10.1007/978-3-662-61184-5_11

M. J. A. M. van Putten, *Dynamics of Neural Networks*,
https://doi.org/10.1007/978-3-662-61184-5_8

consuming processes in the brain. In fact, about 30% of the brain's energy requirement is needed for the processes involved in synaptic transmission. If ischaemia[2] becomes more severe, the sodium-potassium pump will fail, too, and the membrane potential will slowly change,[3] cell swelling occurs (we will discuss the responsible mechanism later in this chapter) and eventually, cell death.

In stroke patients, there is typically a gradient in OGD deprivation: maximally in the center of the infarct zone, with normal values far outside. The region between the central lesion, where typically neurons are irreversibly damaged, and the healthy brain tissue is known as the *penumbra*. In this region, neurons are partially damaged, but may regain (part of) their function, strongly contributing to the rehabilitation process in stroke patients.

8.2 Selective Vulnerability of Neurons

Neurons are selectively vulnerable for oxygen deprivation. In the hippocampus, CA1 pyramidal cells are most sensitive, followed by CA3 cells and dentate granule cells. In the cortex, pyramidal cells in layers III, IV and V are most sensitive and cells in the striatum. Neurons in the brain stem and spinal cord are much more resistant to hypoxic injury [132]. The nature of the neuronal damage varies from reversible synaptic failure to irreversible synaptic failure (that may eventually (time scale days) result in neuronal death) to almost immediate neuronal death (within hours). Also, during and after hypoxia, changes in neurotransmitter release may occur, resulting in excitotoxicity, and synthesis of several proteins, e.g. heat-shock proteins, is altered. The processes that take place in patients with stroke are quite complicated, therefore, with many dynamical interactions. We will discuss some key processes in more detail in the next sections.

8.3 Hypoxia Induces Changes in Synaptic Function

One of the first events that occurs if oxygen supply is limited, is (partial) synaptic failure. Synaptic transmission (we assume chemical synapses) is energetically expensive, estimated to account for approximately 30% of the energy needed by the brain. Initially, synaptic failure is completely reversible, as 'illustrated' in Fig. 8.1. However, if oxygen deprivation lasts longer (minutes or more) synaptic failure will

---

[2]Ischaemia and OGD will be used interchangeably. Although ischaemia means literally deprivation of blood, the resulting effect is insufficient supply of oxygen and glucose.

[3]It is left as an exercise to the student to make an estimate of how fast the membrane potential returns to its new equilibrium value if the sodium-potassium pump is blocked. starting from the normal resting membrane potential. We will discuss later, that the potential difference that is reached is not equal to 0 mV.

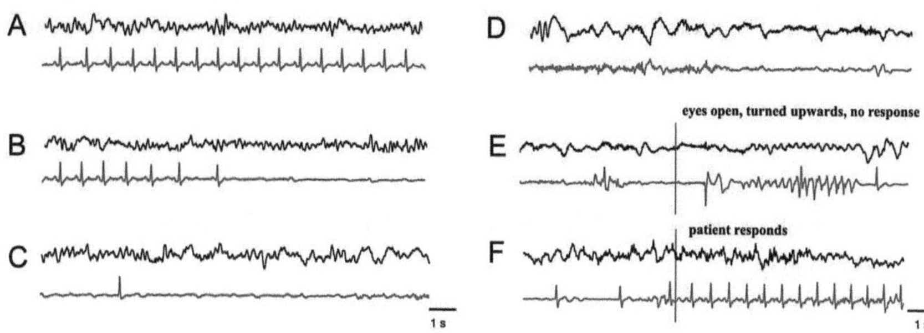

Fig. 8.1 Subsequent 10 s epochs of a single EEG channel and the corresponding EEG recorded during a syncope in a patient who was referred to our department because he experiences recurrent episodes of loss of consciousness. The differential diagnosis included seizures or recurrent syncopes. A. Normal EEG and normal ECG. B. ECG slows down with arrest halfway through the epoch. No change in the EEG. C. Faster EEG activity disappears, and appearance of slow rhythms. D. Further slowing of the EEG, approximately 20 s after arrest. E. Persistent slowing and decrease in amplitude. The patient shows upward deviation of the eyes and is non responsive. F. Full recovery to normal EEG activity within 5 s after the ECG returns. Note that EEG rhythms continue for about 15 s after the heart stopped beating (panel B to beginning of panel D), which illustrates the reserve capacity of ATP. The disappearance of EEG activity mainly results from massive synaptic arrest. This patient was successfully treated with a pacemaker. Illustration from [135]. Reprinted with permission from Wolters Kluwer Health, Inc.

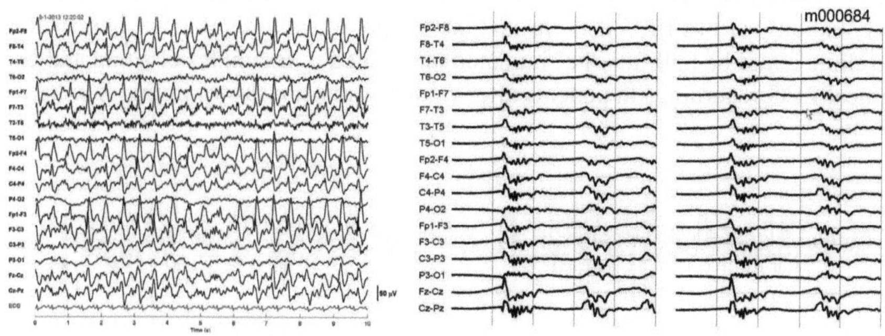

Fig. 8.2 Left: Generalized periodic discharges. This pattern is commonly observed 6–12 h after cardiac arrest and may persist for hours to days. Right: Burst suppression with identical bursts Shown are two epochs of 5 s each, with an interval of 10 s. Note the similarity of the bursts in each channel

last longer, and may be (partially) irreversible. This scenario is often seen in patients after a cardiac arrest. In these patients, cerebral blood flow is critically reduced for many minutes, and neuronal function is severely compromised. Two EEG patterns that can be observed in these patients are generalized periodic discharges and "burst-suppression with identical bursts", illustrated in Fig. 8.2. Both these EEG patterns have a strong correlation with clinical outcome in patients with a postanoxic encephalopathy [54]. These two EEG patterns reflect selective neuronal and synaptic damage, eventually resulting in a reduction of inhibition [53, 93, 122, 134]. Most likely, the glutamatergic synapse that stimulates the inhibitory interneurons is involved. By a reduction of this excitatory input on the inhibitory interneurons, that, on their turn, connect to the pyramidal cells, the overall effect is an overall excitation of the network.

8.4 A Meanfield Model for Selective Synaptic Failure in Hypoxia

In Chap. 7 we discussed a meanfield model with feedback for the EEG, including expressions for the synaptic transmission. Such models can also be used to simulate EEG patterns in pathology that results from changes in synaptic transmission, as in patients with a postanoxic encephalopathy. The model used was based on a meanfield model from David Liley [72], but modified to incorporate time dependent changes in the synaptic transmission. To model the evolution of the EEG after cardiac arrest, it was assumed that briefly after the hypoxic incident, both the inhibitory and excitatory synaptic transmission are affected, represented by excitatory and inihibitory synaptic recovery time constants, τ_e^{rec} and τ_i^{rec}, respectively. Recovery towards baseline values occurs on time scales of hours. To model the effect of long-term potentiation (LTP) of the excitatory synapses, the maximum amplitude of the EPSPs was increased. By choosing plausible values of the short-term time constants and the long-term potentiation, various EEG patterns and the temporal evolution that are observed in the clinic could be faithfully be simulated in the model. Examples of simulated and clinical EEG recordings in patients after cardiac arrest are shown in Fig. 8.3.

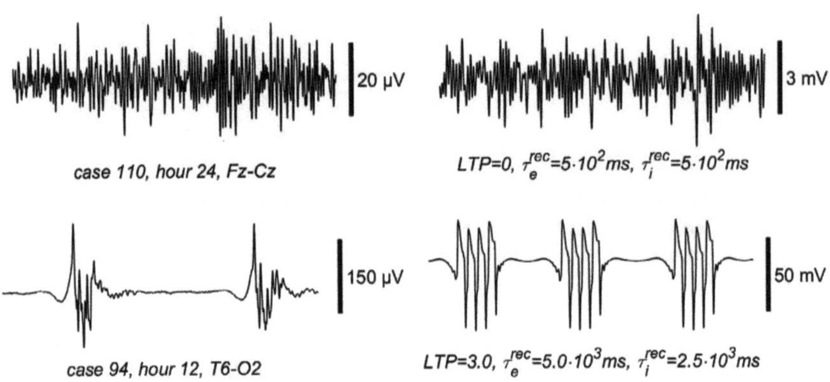

Fig. 8.3 10 s epochs of clinical EEG (left) and simulated (right) EEG data using the model from [94].Top left: normal EEG, 24 h after cardiac arrest. Fz-Cz is the recording position. Bottom left: Burst suppression pattern 12 h after arrest, recorded from T6-O2. Top right: simulated normal EEG; the long term potentiation LTP $= 0$; the excitatory and inhibitory synaptic recovery time constants $\tau_e^{rec} and \tau_i^{rec}$, respectively are indicated. Bottom right: simulation of burst suppression. The LTP of the excitatory synapses is increased. Modified from Fig. 2 in [94], reprinted with permission from Elsevier Ireland Ltd

8.5 Excitotoxicity and Hypoxia Induced Changes in Receptor Function

Both excitatory and inhibitory neurotransmitters are affected during and after hypoxia, which may lead to excitotoxicity. Excitotoxicity is the pathological process by which nerve cells are damaged and killed by excitatory neurotransmitters, for instance glutamate, which is an important excitatory neurotransmitter. Modification of the $GABA_A$ receptor[4] function has also been implicated in several hypoxia-related pathologies, including postanoxic encephalopathy and seizures. Recall that activation of the GABA receptor results from binding of γ-aminobutyric acid (GABA), which regulates the gating of the chloride channel: since $E_{\text{GABA}} < V_{\text{threshold}}$, GABA is an inhibitory neurotransmitter. If hypoxia is more severe, not only synaptic transmission will be affected, but other neuronal processes will fail, too. This will be discussed next.

8.6 The "Wave of Death"

Oxygen and glucose deprivation has almost immediate effects on brain function, typically causing symptoms in approximately 30-40 s, initially resulting in synaptic depression, as discussed in a Sect. 8.3. This dysfunction is also reflected in the electroencephalogram (EEG), generally consisting of an increase in slow wave activity and finally in the cessation of activity (cf. Fig. 8.1). These phenomena are a direct consequence of synaptic failure reflecting the high metabolic demand of synaptic transmission.[5]

Experiments in rats, decapitated to study whether this is a humane method of euthanasia in awake animals, indeed showed disappearance of the EEG signal after approximately 15–20 s. After half a minute of electrocerebral silence, however, a slow wave with a duration of approximately 5–15 s appeared, as illustrated in Fig. 8.4. It was suggested that this wave might reflect the synchronous death of brain neurons [137] and was therefore named the "Wave of Death". To better understand this phenomenon, we modeled the membrane voltage dynamics of a single neuron with a sodium and a potassium channel and leak currents, together with the corresponding changes in the intra- and extracellular ion concentrations, as described in [145]. When a sodium-potassium pump, glial buffering and diffusion of potassium are incorporated to model homeostasis, the model shows regular behavior and has a resting state where all variables obtain values in their physiological ranges. After shutting down the energy supply, the membrane initially depolarizes slowly with a slope of approximately 0.7 mV/s, until it reaches the excitation threshold of the

[4] $GABA_A$ receptors are ligand-gated ion channels, whereas $GABA_B$ receptors are G protein-coupled receptors.

[5] This section rests heavily on our work reported in [145], and part of the text from that paper has been copied to this section.

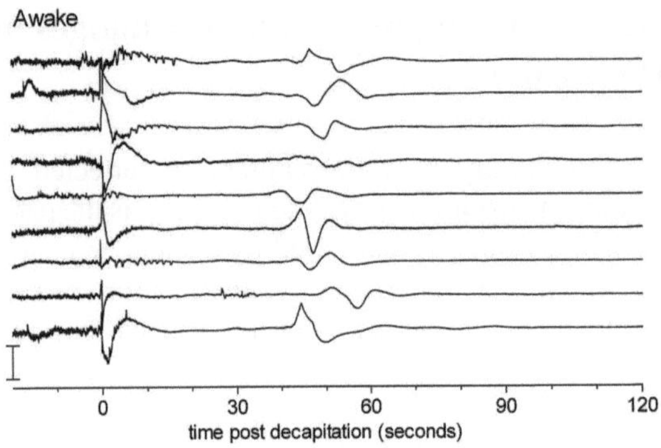

Fig. 8.4 EEGs recorded in 9 rats after decapitation. Each trace represents a one-channel recording from a single animal. Note the large slow wave around 50 s after decapitation. The changes in amplitude at $t = 0$ are movement artifacts due to the decapitation. Figure reprinted from [137]

voltage dependent sodium channel, around 58 mV. Now spiking starts, resulting in an increase in the potassium current with a concomitant reduction in the potassium Nernst potential and membrane voltage. Positive feedback between the increasing firing rate and potassium efflux causes a sudden depolarization of the membrane voltage (30 mV in 2 s), resulting in the membrane depolarization curve, displayed as a dashed line in Fig. 8.5. This behavior was also observed in the in vivo measurements in rats by Siemkowicz and Hansen [99], who also measured a rapid depolarization accompanied with an increase of extracellular potassium, typically 1–2 min after the onset of ischemia. In combination with a high-pass filter, the simulated membrane voltage results in a wave in the EEG as observed by van Rijn et al. (Fig. 8.6, solid line). The "wave of death" therefore, reflects massive neuronal depolarization resulting from complete energy depletion. While modeling the effects of decapitation, an instantaneous cessation of the sodium-potassium pump, glial buffering and diffusion of potassium to the blood was assumed. The last assumption is very reasonable, because arterial pressure vanishes after decapitation, larger vessels are drained and blood flow through the capillaries will stop. The (remaining) blood volume is relatively small and the ion concentrations in the blood will therefore quickly equilibrate with the tissue. However, a complete stop of all active ion transport will not take place directly after decapitation. Some reserves of metabolic substrates and ATP are still left in the tissue. In human brain tissue for example, these reserves can support a maximum of one minute of normal metabolism, but less if no oxygen is available. Such effects do not disqualify the general behavior of the model, as they will only result in a delay in the onset of depolarization.

A single neuron model was used to calculate an EEG. Although usually the network properties of neurons are essential for the EEG, we argued that a single neuron approach is realistic because synaptic transmission ceases quickly during anoxia and neurons therefore no longer receive input. Although the postsynaptic response is still intact, for example the response of the neuron to glutamate [14], neurotransmitters

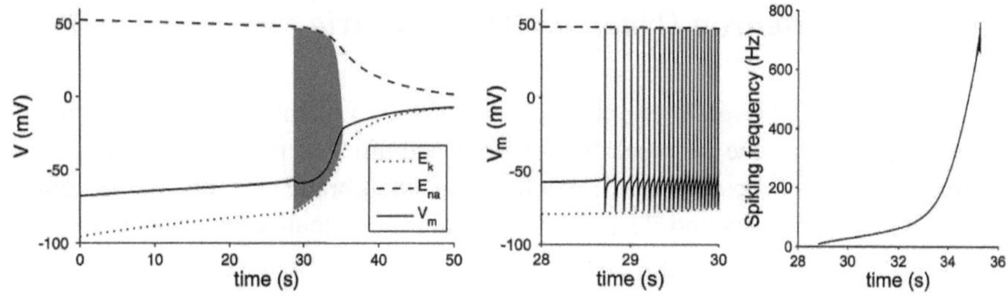

Fig. 8.5 Membrane dynamics during oxygen-glucose deprivation. In the left panel the membrane dynamics are shown that occur after the onset of OGD (solid line). The dashed and dotted lines show the progressive loss of ion gradients. When after a gradual rise the membrane potential reaches the excitation threshold, this subsequently results in spiking of the membrane voltage (gray region, not resolved). The black line shows the average membrane potential during the spiking (averaged over 300 ms). After approximately 7 s of oscillations, the cell comes to rest again, with a resulting membrane potential at the Donnan equilibrium (to be discussed next). The middle panel shows a close up of the start of spiking activity, the right panel shows the instantaneous firing rate. Reprinted from [145]

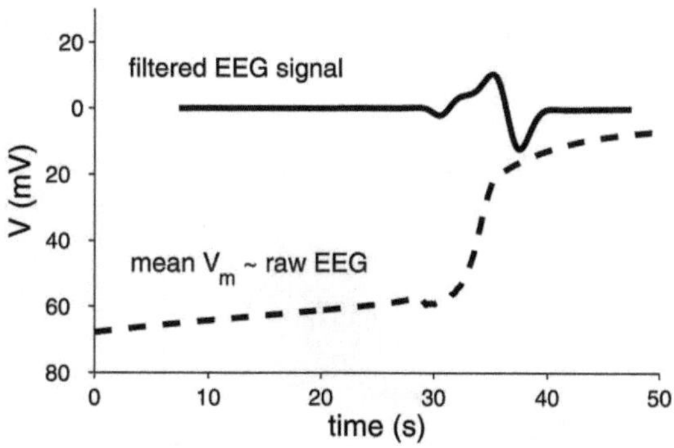

Fig. 8.6 Mean membrane potential and simulated EEG signal. Shown are (dashed line) the simulated membrane potential averaged over 300 ms and (solid line, a.u.) the signal that results after applying a high-pass filter (second order Butterworth filter, cut-off at 0.1 Hz). Reprinted from [145]

are no longer released and transmission is halted. The absence of significant EEG power after about 20 s post decapitation as observed by van Rijn et al. [137] most likely results from this failure of synaptic transmission. The depolarization wave was observed during a relatively short period of several seconds. As the extracellular currents generated by a single pyramidal neuron are of the order of pA, much too small to generate a measurable scalp potential, a very large number of cortical neurons must simultaneously depolarize after decapitation. More details about this model can be found in [145].

8.6.1 Single Neuron Dynamics After Hypoxia

Neurons that are deprived of energy will not be able to maintain their resting membrane potentials: as the permeability of both sodium and potassium is non-zero, the ion gradients of these ions will result in transmembrane currents that generally reduce the ion gradients and decrease the membrane potential. Neurons will spontaneously spike as the voltage-gated sodium channels are activated. However, in experiments with single neurons using blockers of the sodium-potassium pump, this spiking behaviour appears to be rather diverse: some neurons only spike once, while others show spike trains in the transition to the depolarized state, as illustrated in Fig. 8.7. On closer inspection, this can be understood using the following reasoning: although sodium and potassium ion gradients will be reduced after blockage of the sodium-potassium pump, it is highly unlikely that the time course of the gradients will be the same for different neurons. Indeed, due to biological variations in e.g. the size of the extracellular space and pump densities, this will vary. If this variation in the temporal behaviour of the change in Nernst potentials of sodium and potassium is considered, a bifurcation diagram of the transitions from resting membrane potential towards the emergence of the depolarization block (the new equilibrium) can be

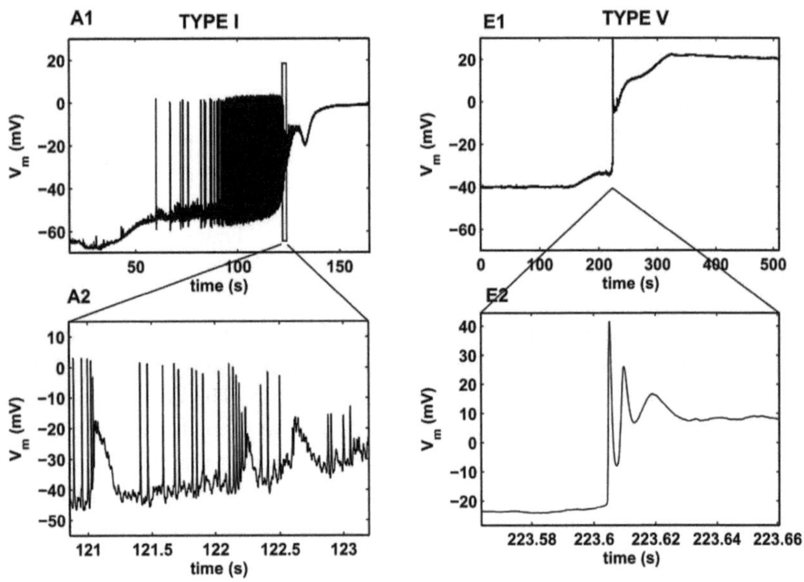

Fig. 8.7 Time course of the membrane voltage after 30 s perfusion with 200 μM ouabain to block the sodium-potassium pump. Two examples of spontaneously spiking neurons (pyramidal cells) are shown. Left: repetitive spiking. Right: single spike and relaxation oscillation from a polarized to a depolarized state. Eventually both neurons reach a "depolarization block-like state". Illustration from [144]

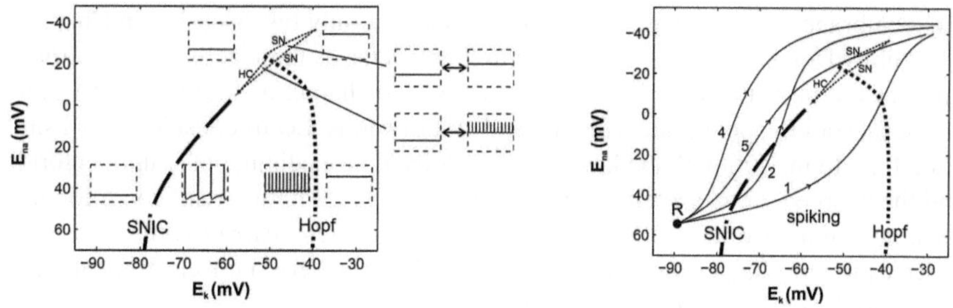

Fig. 8.8 Left: bifurcation diagram of a Hodgkin-Huxley (HH) type model. Shown are the saddle-node-on-invariant- circle (SNIC) bifurcation (thick dashed line), which corresponds to a class 1 spiking threshold, and the Hopf bifurcation (thick dotted line), which corresponds to a depolarization block A saddle-homoclinic orbit (HC) and 2 saddle-node (SN) bifurcations (thin dashed lines) connect the SNIC and Hopf lines. Insets: qualitative dynamics of the membrane voltage at different positions in the diagram. In the small area between the HC and SN bifurcation lines, the neuron dynamics are bistable (denoted by double-headed arrows). E_{Na} and E_{K} are the Nernst potentials of sodium and potassium, respectively. Bifurcation lines are calculated from the diagram depicted in (Barreto and Cressman 2011 [8]). Right: Trajectories through the bifurcation diagram of a Hodgkin-Huxley type model. Physiological resting state is denoted by R. Solid lines denote 4 hypothetical trajectories of the ion concentrations, leading to the experimentally observed dynamics. Trajectory 3 (not shown for clarity) lies slightly below trajectory 5 and crosses the HC bifurcation. Illustration from [144]

sketched as shown in Fig. 8.8. All these possible transitions were indeed experimentally observed. Another phenomenon that will occur in severe energy deprivation is cell swelling, or *cytotoxic edema*, to be discussed next.

8.7 The Gibbs-Donnan Effect and Cell Swelling

Why do cells eventually swell if their energy supply is severely limited? Recall from our discussion of the Nernst potential in Chap. 1, Sect. 1.2.2 and the previous Sect. 8.6, that we needed the Na/K-ATP pump to compensate for the leak currents, as otherwise ion concentration gradients would eventually be lost. In reality, however, if we wait sufficiently long after the ATP dependent Na/K pump has stopped, ion gradients will not become zero. This is caused by the fact that the inside of the cell contains many large negatively charged proteins, that cannot pass the membrane. We will also discuss that the presence of these non-permeable negatively charged proteins results in significant changes in osmotic pressure as the pumps are halted, resulting in cell swelling or *cytotoxic edema*.

Charged particles may fail to distribute evenly across a semipermeable membrane if there is at least one charged substance that is unable to pass the membrane while the membrane is semi-permeable for *more than one ion species*. This phenomenon is known as the *Gibbs-Donnan* effect. Note, that this is different from a situation where the membrane is permeable for a *single* ion species, only (as discussed in Chap. 2). The Gibbs-Donnan effect is ubiquitous in living cells, as these cells contain various impermeant huge macromolecules with a negative charge, in addition to the small permeant sodium, potassium, calcium and chloride ions. Let us consider such a situation for a living cell with the key players involved, i.e. sodium, potassium, chloride and the charged, large proteins (with negative charge). An important condition that must be maintained in a solution with ions is *bulk electroneutrality*: the sum of all the free moving charges must equal zero.[6] In order to preserve bulk electroneutrality, it thus holds for the extracellular space that

$$[Na^{+}]_e + [K^{+}]_e - [Cl^{-}]_e = 0 \tag{8.1}$$

with $[X]_e$ the extracellular concentration of ion species X. Similar, to preserve bulk electroneutrality in the intracellular compartment, it is required that

$$[Na^{+}]_i + [K^{+}]_i - [Cl^{-}]_i + \rho_{\text{macro}}/e = 0, \tag{8.2}$$

with ρ_{macro} the total charge density of the impermeant macromolecules. Another condition that needs to be fulfilled is that in equilibrium each *permeant* species must be in Nernst equilibrium at the same value of the membrane potential, V_m. Recall, that impermeant ions do not *primarily define* the membrane potential, as charge separation can only occur if an ion can *move across* the cell membrane. This is why the impermeant macromolecules are not present in the expression that defines the equilibrium membrane potential if all energy dependent processes are halted, given by:

$$V_m = -\frac{k_b T}{e}\ln\frac{[Na^{+}]_e}{[Na^{+}]_i} = -\frac{k_b T}{e}\ln\frac{[K^{+}]_e}{[K^{+}]_i} = \frac{k_b T}{e}\ln\frac{[Cl^{-}]_e}{[Cl^{-}]_i}. \tag{8.3}$$

Rewriting this equation as the Gibbs-Donnan relation, we have (in equilibrium)

$$\frac{[Na^{+}]_i}{[Na^{+}]_e} = \frac{[K^{+}]_i}{[K^{+}]_e} = \frac{[Cl^{-}]_e}{[Cl^{-}]_i}. \tag{8.4}$$

In this equilibrium condition, all available permeant ions essentially "share in the job of neutralizing the huge macromolecules with negative charge, ρ_{macro}" [80].

8.7.1 Calculation of Gibbs-Donnan Potential

Let us assume a model cell placed in an infinite volume with concentrations given by Table 8.1. Let us now calculate the ion concentrations inside our model cell that

[6]Note, that this does not exclude charge separation across the semipermeable membrane as these charges do not freely move in the bulk solution.

Table 8.1 Values for the ion concentrations the extracellular space and the equivalent mM of electrons from the macromolecules, that are intracellular, only. Note, that the solution of the extracellular compartment is indeed electrically neutral. The charged macromolecules that cannot permeate the membrane (e.g. large negatively charged proteins) are indicated with ρ_{macro} with an equivalent of 125 mM electrons

Ion species	Concentration (mM)
$[Na^+]_e$	140
$[K^+]_e$	10
$[Cl^-]_e$	150
ρ_{macro}	125

will be present at the Gibbs-Donnan equilibrium. If we set $[Na^+]_i = x$, it follows that

$$\left(1 + \frac{0.01}{0.14}\right)x^2 - 0.15 \cdot 0.14 - 0.125x = 0. \tag{8.5}$$

Solving for x results in $x = 0.21$ or $[Na^+]_i = 210$ mM. From (8.3) we now find that the membrane potential is -10 mV. Check that this also holds if you use the potassium or chloride concentrations (you should first find that $[K^+]_i = 15$ mM and $[Cl^-]_i = 100$ mM).

Check that the former result follows straightforward from combining (8.2) and (8.4), that can be expressed as:

$$[Na^+]_i + \frac{[K^+]_e[Na^+]_i}{[Na^+]_e} - \frac{[Na^+]_e[Cl^+]_e}{[Na^+]_i} + \rho_{\text{macro}}/e = 0 \tag{8.6}$$

and subsequently multiplying with $[Na^+]_i$.

Even if the initial concentrations of the cations are the same, as illustrated in Fig. 8.9, a membrane potential will arise that is different from zero, resulting from the Gibbs-Donnan equilibrium. In this example, the osmotic pressures are also similar, as the total amount of ions in solution on either side is identical. It is apparent that this situation cannot be an equilibrium state: since the membrane potential is zero, there acts no electrical force on the ions, and the concentration gradient will drive the extracellular ions B^- into the cell. However, bulk electroneutrality must be preserved, which is realized by "dragging along" positively charged ions, A^+ to the interior as well (even against their concentration gradient!).

The permeable anions will not bring along to the interior all the positively charged ions from the exterior compartment, and a fraction of the positive charge will remain in the exterior volume. But these positive charges cannot remain in the bulk solution in the external compartment, as this would offend bulk electroneutrality, and will therefore also reside near the membrane, but now on the opposite site. So the actual charge *separation* consists of negatively charged ions internally and positively charged ions at the exterior, both in very close contact to the membrane. In

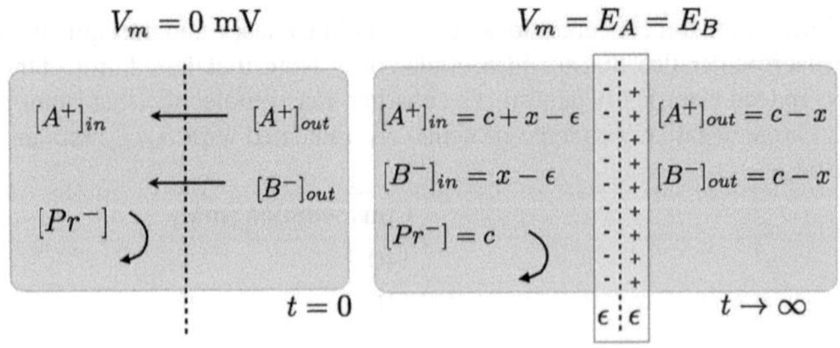

Fig. 8.9 Cartoon of ion gradients and possible fluxes in the calculation of the membrane potential resulting from the Gibbs-Donnan effect. In this situation, the membrane is semipermeable for both ion species A and B. Left panel: initial condition, the system is not in equilibrium. Right panel: in equilibrium, bulk electroneutrality is preserved; at the same time, a very small fraction of the cations and anions (ε) is separated from the bulk and resides on each side of the membrane, creating a potential difference. This equilibrium condition is very different from the situation where the membrane is permeable to a single ion species, only. The associated ion fluxes x in reaching the Gibbs-Donnan equilibrium are a significant fraction of the initial concentrations c. The negatively charged proteins are indicated with $[Pr^-]$, assumed to be present in the intracellular compartment, only

equilibrium, the electrical field resulting from this charge separation will counteract further diffusion of both A^+ and B^-. This is schematically illustrated in Fig. 8.9, right-panel.

The equilibrium state is sketched in the right of Fig. 8.9: The concentrations of the permeable ions are such that their Nernst potentials E_A and E_B are both equal to the membrane potential V_m, which is proportional to the charge separation ΔQ.

This yields two equations for the two unknowns x, the difference between the initial and equilibrium value of the extracellular concentrations of A^+ and B^-, and ε, which is a measure for the ions bound to the membrane and proportional to ΔQ. Note, that the change in the bulk concentration of ions in the extracellular compartment in this example is $-x$, but $x - \varepsilon$ is actually available in the intracellular compartment, as a charge of amount ε remains in the fictitious compartment near the cell-membrane. In equilibrium it holds that all Nernst potentials need to be equal, thus we can write for the ion concentrations $[A^+]$ and $[B^-]$

$$-62\log_{10}\left(\frac{c+x-\varepsilon}{c-x}\right)\text{mV} = +62\log_{10}\left(\frac{x-\varepsilon}{c-x}\right)\text{mV}. \tag{8.7}$$

As we will show in the Exercises, $\varepsilon \ll x$, and the Gibbs-Donnan equilibrium concentrations are now obtained by solving

$$\frac{c-x}{c+x} = \frac{x}{c-x} \tag{8.8}$$

that results in $x = c/3$. At equilibrium, the membrane potential difference is therefore

$$-62\log_{10}(2) \approx -18.7 \text{ mV}. \tag{8.9}$$

8.7.2 Cell Swelling

If we consider the sodium gradient and its associated Nernst potential in the squid giant axon, we find that $V_{\text{Nernst,Na}} = +54$ mV, while the Nernst potentials of potassium (-75 mV) and chloride (-59 mV) are near the resting membrane potential of the squid axon, about -60 mV i.e. both potassium and chloride are nearly at equilibrium. As the Nernst potential of the various ions, in particular sodium, are different from the membrane potential, sodium ions will diffuse to the intracellular compartment and potassium ions in the reverse direction, where their currents are given by $g_i(V_m - E_i)$, with $i \in$\{Na,K,Cl\}. As these currents are generally non-zero, there must be an active mechanism involved to maintain the sodium gradient, and keep the cell from reaching the Gibbs-Donnan equilibrium, as was also discussed in Chap. 1. This is realized by the sodium-potassium pump, that actively exchanges intracellular sodium with extracellular potassium, using ATP as the fuel.

This process fails in patients with cerebral infarcts. In that situation, the sodium-potassium pump is severely compromised. In the core area of the infarct, ATP cannot longer be generated by the mitochondria, being completely deprived of oxygen and glucose, and the sodium-potassium pump can no longer maintain the ionic gradients. Neurons will therefore reach their Gibbs-Donnan equilibrium, and changes in osmotic pressure result from the large ionic fluxes to the intracellular compartment. This increase in osmotic pressure will result in swelling of neurons, and soon after neuronal death.

In severe cases, this can result in a cascade of progressive cell swelling, as was illustrated in Fig. 1. This scenario of *malignant edema* most likely depends on e.g. the ATP gradient from the core to the peri-infarct brain tissue (the penumbra[7]), e.g. the size of the infarct core and the amount of 'space' available in the neurocranium, as an elevated intracranial pressure will cause collateral damage from a reduction in blood flow and mechanical pressure on the cell membranes. In one of the Exercises, you will make an estimate of the osmotic pressure associated with the Gibbs-Donnan equilibrium.[8]

---

[7]In the penumbra, neuronal function is severely compromised, but there is potential for (some) recovery.

[8]Plant cells cope with the osmotic pressure resulting from the Gibbs-Donnan equilibrium by a rigid wall and turgor pressure.

8.7.3 Critical Transitions in Cell Swelling

If energy is significantly deprived and ion gradients cannot be maintained, cells will eventually swell. The temporal dynamics depends on various conditions, including membrane permeabilities and remaining pump strengths. In [35], a relatively simple model of a single cell in an infinite extracellular space was introduced to model this process, illustrated in Fig. 8.10. It was observed in the simulations that if the sodium pump strength is gradually decreased, a bifurcation occurs after which the cell membrane potential changes to a pathological equilibrium, and cell swelling starts, as illustrated in Fig. 8.11. Note further that to bring the neuron back to its physiological state, the pump strength has to increase to values far above 150% of its resting value, which may preclude return as such strengths may not be feasible. Indeed, it has been experimentally observed that neurons can remain in their depolarized state after restoration of energy [18].

8.8 Spreading Depression

Spreading depression (SD), or depolarization, is a phenomenon characterized by a slowly traveling wave (mm/min) of neuronal depolarization with a concomitant redistribution of ions between the intra- and extracellular space. Spreading depression was

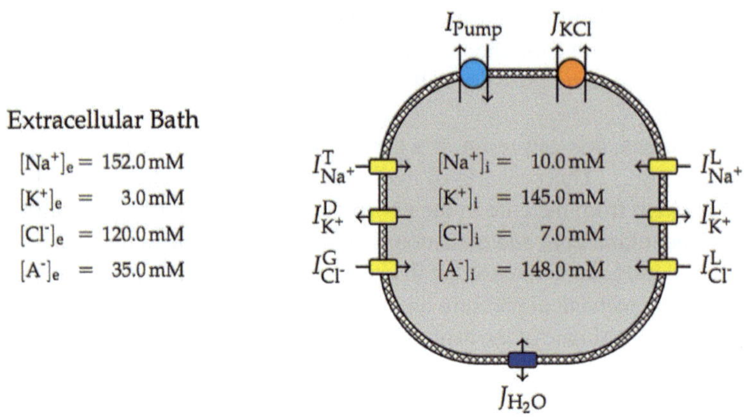

Fig. 8.10 Model of a single neuron with typical ions concentrations. Negatively charged, impermeant macromolecules are denoted by A^-. Leak and voltage-gated ion channels (yellow) yield ion currents that are balanced by the electrogenic ATP-pump (cyan) and the electroneutral KCl cotransporter (orange). While the pump moves both Na^+ and K^+ against their electrochemical gradients and therefore needs ATP to run, the KCl cotransporter uses the energy stored in the transmembrane gradient of K^+ to move Cl^- out of the cell. Any difference in osmolarity between the intracellular and extracellular space will yield a water flux across the membrane (blue), changing the cell volume. Illustration from [35]

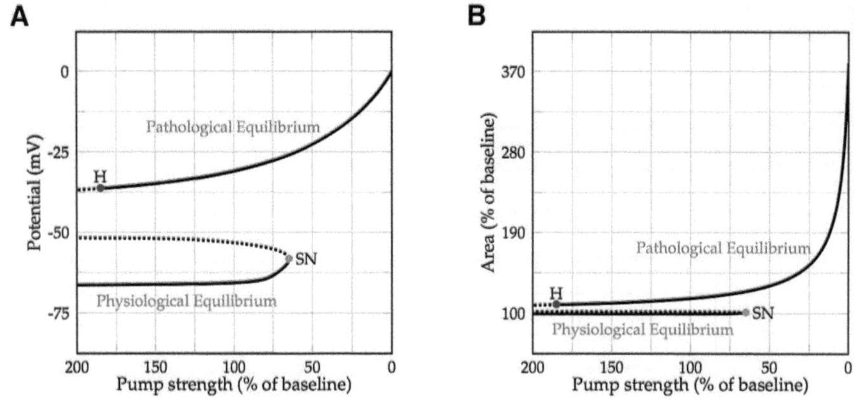

Fig. 8.11 Bifurcation diagrams with the Na/K-ATPase strength as a free parameter. **a** stable equilibria are denoted by a solid line and unstable equilibria by a dotted line. At approximately 65% of the baseline pump strength, the physiological resting state disappears via a saddle node bifurcation (SN, orange). For lower values of the pump strength the cell will converge to a depolarized Gibbs-Donnan-like equilibrium. This pathological state is stable for pump strengths of up to approximately 185% of the baseline pump rate, where it loses stability due to a subcritical Hopf bifurcation (H; blue). **b** The cell volume is almost constant in the physiological equilibrium branch, but is highly dependent on pump strength in the pathological equilibrium branch, where minor differences in the remaining pump rate cause major differences in equilibrium cell size. Illustration from [35]

discovered by Aristides Leão, published in 1944 [70]. He discovered that spontaneous electrical activity can be depressed in response to electrical stimulation of cortical tissue from the rabbit, illustrated in Fig. 8.12. The records show that "the depression spread out slowly, in all directions, from the region stimulated...". After full suppression, activity slowly recovers, starting at the site that was initially suppressed. He also observed that mechanical stimulation could induce a spreading depression and that particular drugs (strychnine, acetylcholine, eserine) could also induce a SD. At the time of the original observation, details of the processes responsible for the depression of activity were not yet known.

It is now understood that ion redistributions are responsible for the temporary depression of electrical activity [25, 147] and that SD can be induced by many stimuli, including ischemia, electrical stimulation, mechanical damage (needle prick), or application of potassium or or glutamate. All these stimuli directly or indirectly increase neuronal excitability or depolarize neuronal membranes. If an SD is triggered, it propagates in an all or none fashion in all directions, independent of the stimulus type or intensity, similar to an action potential. Several hypotheses have been put forward to explain the propagation of SD. Most likely, diffusion of potassium and glutamate through the extracellular space cause the depression of neuronal activity, but other mechanisms may be involved, too (e.g. the effect of gap junctions). In physiological conditions, many control mechanisms are involved in the homeostasis of potassium and glutamate. Key components in this control include the pre-and

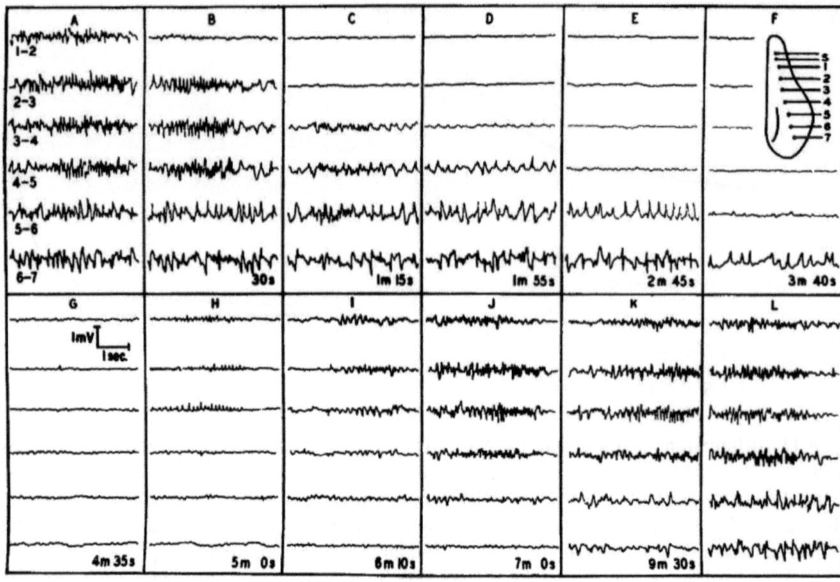

Fig. 8.12 Gradual spread of depression in a slice of the cerebral cortex of a rabbit. The electrical stimulation is applied at the electrode labeled S. Electrodes arranged as shown in the inset. A. Before stimulation. Panel L is recorded 7 min. after panel K. Reprinted from [70] with permission from the The American Physiological Society

postsynaptic neuron, glia and the vasculature, known as the neurovascular unit [58, 82], illustrated in Fig. 8.13 Recall that the resting membrane potential, given by (1.10), discussed in Chap. 1, and repeated for convenience is given by

$$V_{\text{rest}} = \frac{g_{\text{Na}} E_{\text{Na}} + g_{\text{K}} E_{\text{K}} + g_{\text{Ca}} E_{\text{Ca}} + g_{\text{Cl}} E_{\text{Cl}}}{g_{\text{Na}} + g_{\text{K}} + g_{\text{Ca}} + g_{\text{Cl}}}. \tag{8.10}$$

It follows directly that membranes can be depolarized by either changing the Nernst potentials or the conductances. While fast changes are mediated by changes in gating, the sustained depolarizations and slow membrane dynamics observed during SD can result from changes in the resting membrane voltage mediated by changes in the intra- and extracellular ion concentrations. The dynamics of the spread of the SD can be described by a reaction-diffusion equation [146]. While diffusion from a fixed source becomes progressively slower over longer distances, a reaction-diffusion process (propagates at a steady velocity by recruiting medium (tissue) at the front of the wave as new source. For an idealized case, the velocity is given as [146]

$$v = \sqrt{\frac{R D_{\text{eff}}}{\Delta C}} \tag{8.11}$$

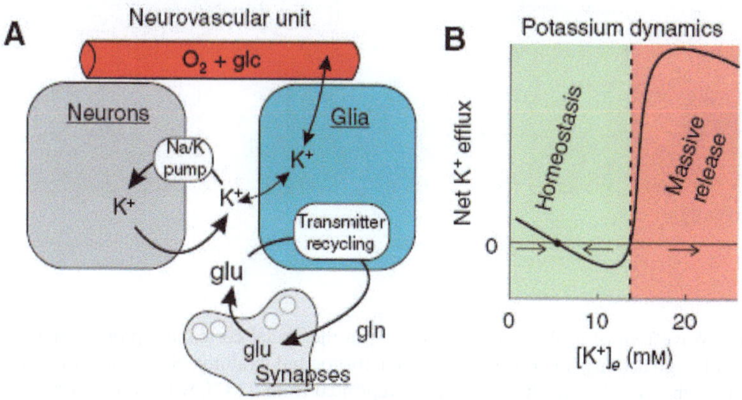

Fig. 8.13 Homeostasis of extracellular potassium and glutamate in the neurovascular unit. Panel A schematically shows the main release and uptake pathways. Potassium leaking from the neurons and released during action potentials is pumped back by the Na/K-pump. Released glutamate (glu) is taken up by glia cells and returned in the form of glutamine (gln). In addition, glia can rapidly buffer K^+, distribute it over the glial syncytium, and transport it to the blood stream. A constant supply of oxygen and glucose from the blood is necessary to fuel these processes. Panel B shows a sketch of the dynamics of extracellular potassium. Up to a threshold (dashed line) of typically 8–20 mM, elevated extracellular potassium increases its own removal from the extracellular space by stimulating Na/K-pumps and glial uptake. This restores the concentration to the physiological set point (black dot). Above threshold (dashed line), potassium is released into the extracellular space faster than its removal due to stimulation of neuronal action potential generation. The dynamics of extracellular glutamate show a similar threshold (not shown). Reprinted from [147], with permission from Walter de Gruyter publishers

where R is the rate at which neurons expulse potassium or glutamate, D_{eff} is the effective diffusion constant, and ΔC is the concentration threshold above which neurons start expulsing this substance. As potassium and glutamate are charged, they cannot diffuse freely: their charges must be balanced by other ion movements with opposite charge to preserve electroneutrality. This is one of the reasons that detailed computational modeling of SD is not trivial [25, 146].

SD occur in many neurological conditions, such as migraine with aura [28, 29], ischemic stroke [56], traumatic brain injury [69] and possibly epilepsy [37, 38, 69, 103, 147]. This can be clinically observed, too: I have seen several patients with migraine with aura that report that their migraine attacks can start with positive phenomena (e.g. light flashes in the right visual field), followed by partial loss of vision, and 3–5 min later by sensory disturbances in their right arm, starting at the hand area, and gradually moving towards the shoulder, in some cases (several min later) followed by problems with language understanding (receptive dysphasia); if asked, visual problems are then typically gone. After about 5–10 min, headache may start, and language function has recovered. This is the clinic of a "wave" that starts at the occipital region, and slowly moves anteriorly towards the sensory cortex, subsequently to the motor strip and eventually towards Wernicke. Such a spread of depolarization is illustrated in Fig. 8.14.

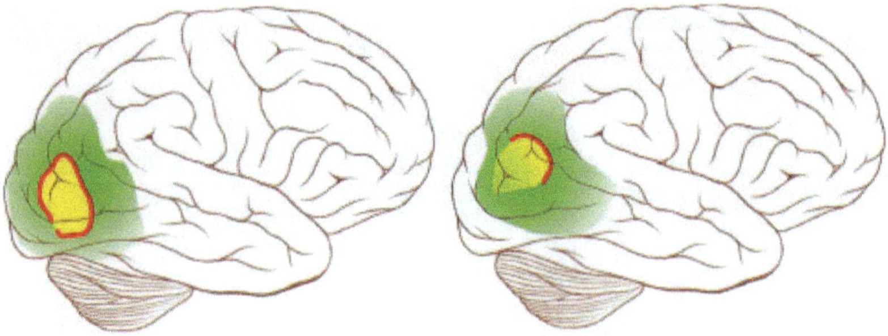

Fig. 8.14 Wavefront of a SD (red line), starting in the visual cortex and propagating anteriorly. Colors mark high activity (or concentration) of activator and inhibitor. Note that in the panel on the right, part of the visual cortex has recovered. Reprinted from [147], with permission from Walter de Gruyter publishers

If energy supply is not limited, the disturbed ion gradients typically recover after a few minute after SD onset and function recovers. A critical factor is that the Na/K-pump can restore the extracellular potassium. Failure to do so will result in irreversible neuronal dysfunction. This latter process is involved in the peri-infarct depolarizations, that can contribute to irreversible damage of penumbral neurons.

8.9 Clinical Challenges

Stroke is a common clinical condition, with a yearly incidence of approximately 30.000 patients in the Netherlands. Although techniques to treat patients have significantly improved, including the use of intra-arterial cloth removal, there are still many patients for which this treatment is not possible, for instance due to technical limitations or significant delay in hospital admittance. A better understanding of processes that are critically involved in recovery of penumbral neurons may result in better treatments.

In patients who are successfully resuscitated, many do not immediately regain consciousness, and are treated in the intensive care unit. In the Netherlands, this amounts to approximately 7000 patients per year. Important clinical questions include: what is the optimal treatment for these patients? Can we define reliable prognostic factors for neurological outcome? What is the optimal rehabilitation strategy? Clinical and fundamental research in this will benefit from a good understanding of some basic pathophysiological mechanisms.

8.10 Summary

We discussed why neurons need energy and various dynamical processes that occur after energy deprivation, including changes in membrane potential and cytotoxic edema. You learned which processes are critical in the development of cytotoxic edema and how this can be illustrated with bifurcation diagrams. We also discussed spreading depression, and its relevance in e.g. migraine and stroke.

Problems

8.1 What potential mechanisms can be involved in the selective vulnerability of neurons to hypoxic incidents?

8.2 Bolay stated in one of her papers [14] that in some patients with brain infarcts, the MRI may not show any abnormality. This was also discussed in [55]. What are the arguments put forward in these papers?

8.3 Which neuroransmitters and other substances can interact with the $GABA_A$ receptor?

8.4 Show that for realistic ion concentrations, ε is indeed negligible, as assumed in the derivation of the Gibbs-Donnan equilibrium. Assume a spherical cell with radius of 10 μm and use for the specific membrane capacity $c_m = 0.01$ F/m^2 and take ion concentrations as illustrated in Fig. 8.9.

a. Estimate the total membrane capacitance of the cell.
b. Calculate the charge accumulated across the membrane that is needed to generate the resting membrane potential. Use initial concentrations as shown in Fig. 8.9 and take the assumption that ε can be neglected.
c. Now show that you may assume that ε is very small by estimating the total charge inside this spherical cell. You can set c = 120 mmol/liter (c = 120 mM).

8.5 The shift in ions that occur if the Gibbs-Donnan equilibrium is obtained results in a very large osmotic disbalance. This will result in water entering the cell, eventually destroying the cell membrane. Typical concentration differences of a biological cell associated with the Gibbs-Donnan potential are about 10–20 mM, where M denotes mole per liter.

a. Can you now make an estimate of the increase of osmotic pressure?
b. What is the primary mechanism that prevents this process from happening?

8.6 In this exercise,[9] we estimate the time it takes to reach the Gibbs-Donnan equilibrium for the squid giant axon if we stop the sodium-potassium pump. Assume

[9]Based on Exercise 11.4 in Philip Nelson, Biological Physics, W.H. Freeman and Company, 2008

Table 8.2 Ion concentrations across a semipermeable membrane for Problem 8.7

Intracellular (mmol/l)	Extracellular (mmol/l)
$Cl^- = 50$	$Cl^- = 149.9$
$K^+ = 150$	$K^+ = 150$
$A^- = 100$	$A^- = 0.1$

for the specific membrane conductance $g = 0.2$ S/m^2 and for the relative conductances $g_K = 25g_{Na} = 2g_{Cl}$. In this exercise, we will assume that all conductances are constant.

a. What are the specific membrane conductances of potassium, sodium and chloride?
b. Assume the diameter of the axon is 1 mm. Take for the extracellular sodium concentration 440 mmol/l, the intracellular Na-concentration 50 mmol/l, intracellular potassium 400 mmol/l, extracellular potassium 20 mmol/l, intracellular chloride 52 mmol/l and extracellular chloride 560 mmol/l. Further assume that the squid is at a temperature of 7 °C. What is the initial sodium current immediately after the Na-K pump is switched off?
c. What is the total charge carried by the sodium ions inside the axon per unit length?
d. Estimate the time it takes to reach an intracellular sodium concentration of 100 mmol/l, assuming that V_m remains constant.

8.7 Consider a system with a semipermeable membrane that allows passage of chloride and potassium, but not of macromolecules with negative charge, A^-, with concentrations (mmol/l) given in Table 8.2. If Gibbs-Donnan equilibrium is reached, what will be the value of the ion concentrations across the membrane?

8.8 Consider a neuron immersed in an infinite extracellular solution with $[Na^+]$, $[K^+]$, $[Cl^-]$ ions (constant concentrations). The same ions are present in the intracellular solution. In addition, negatively charged macromolecules, $[A^-]$ are present, both intra- and extracellular. Show that the intraneuronal sodium concentration at GD equilibrium can be written as a function of the extracellular cations and $[A^-]$, only, i.e. $[Na^+]_i = f([Na^+]_e, [K^+]_e, [A^-]_i, [A^-]_e)$ with i, e the intra- and extracellular solution, respectively.

Chapter 9
Seizures and Epilepsy

People think that epilepsy is divine simply because they don't have any idea what causes epilepsy. But I believe that someday we will understand what causes epilepsy, and at that moment, we will cease to believe that it's divine. And so it is with everything in the universe.

— Hippocrates

Abstract In this chapter, we present clinical aspects of seizures and epilepsy. We discuss how dynamical models can advance our understanding of epilepsy and seizures, and introduce candidate targets for treatment. Pharmacoresistance is shortly discussed.

9.1 Introduction

Epilepsy is a common neurological disorder with a prevalence of approximately 1%. Epilepsy can be defined as a brain condition with an increased likelihood of *seizures*. Seizures are changes in behaviour, experience or motor function due to abnormal, typically excessive, synchronization between neuronal populations [130, 139]. This pathological state may be relatively limited in spatial distribution (focal seizures) or can involve both hemispheres (generalized seizures). Such increased synchrony is, in most patients, also observed in the electroencephalogram during a seizure. An example was shown in Fig. 3.1. Clinical manifestations are extremely varied, ranging from subtle changes in perception, for instance smelling a particular substance or perceiving fear to generalized tonic-clonic seizures accompanied by loss of consciousness. After the end of the seizure, the postictal state starts, that can be defined as "a temporary brain condition following seizures (a) manifesting neurological deficits and/or psychiatric symptoms, (b) often accompanied by EEG slowing or suppression,

Electronic supplementary material The online version of this chapter (https://doi.org/10.1007/978-3-662-61184-5_9) contains supplementary material, which is available to authorized users.

The original version of this chapter was revised: The belated corrections have been incorporated. The correction to this chapter is available at https://doi.org/10.1007/978-3-662-61184-5_11

M. J. A. M. van Putten, *Dynamics of Neural Networks*,
https://doi.org/10.1007/978-3-662-61184-5_9

(c) lasting minutes to days" [87]. Postictal symptoms include coma, paresis (Todd's paresis), dysphasia and delirium.

Epilepsy may result from a genetic or predisposition for seizures, for instance channelopathies or inborn errors of metabolism. Other congenital epilepsy syndromes include migration disorders or tuberous sclerosis. Many other forms of epilepsy are acquired. Almost all types of brain injury, ranging from hemorrhagic or ischemic stroke, global ischaemia, traumatic brain injury and inflammation, can result in seizures and epilepsy, in particular if the cortex is involved. In patients with acute stroke, the peri-infarct cortex is generally considered to be the seizure generating zone. Acute seizures in patients with brain injury are often transient, e.g. resulting from an increase in extracellular potassium caused by neuronal injury, while chronic and persistent seizures may result from permanent changes in functional connections, for instance resulting from changes in synaptic efficacies. As the cortex is intimately connected to the thalamus, primary insults to the cortex may also induce changes in the corticothalamic circuitry, resulting in a persistent increase in the likelihood of generating seizures. This candidate mechanism is suggested by a recent study in rats, where it was shown that a cortical stroke can enhance the excitability of thalamocortical cells, causing seizures [84].

The processes that transform normal neuronal circuits into epileptic circuits is known as *epileptogenesis* and the mechanisms that generate seizures as *ictogenesis*. In general, both are still only partially understood. We know many associations between particular clinical conditions and epileptogenesis and ictogenesis. For instance, patients after a stroke [41], traumatic brain injury or encephalitis have in increased risk of seizures. These are examples of acquired epilepsies. Several channelopathies, often resulting from a genetic disorder, may result in seizures and epilepsy, too. For instance, in benign familial neonatal convulsions type 2 an abnormal α-subunit of a voltage-gated potassium channel (KCNQ3, 8q24) has been identified, in some forms of juvenile myoclonic epilepsy, the voltage-gated chloride channel (CLCN2, 3q27.1) may be involved. Ligand-gated channels may also be affected. For instance, in autosomal dominant juvenile myoclonic epilepsy, changes in the ligand-gated GABA-A receptor, α-1 subunit (GABRA1, 5q34–q35) have been identified. However, the fundamental mechanisms involved in the generation of the seizures is ill understood. We have only a limited understanding of why and how particular channelopathies result in abnormal neuronal synchronization and why many patients after traumatic brain injury or stroke develop epilepsy.

9.2 Prevention and Treatment of Seizures

In approximately 70% of patients seizures can adequately be prevented with medication, only. Most of these anti-epileptic drugs modify voltage-gated or ligand-gated receptors. Examples include sodium valproic acid and phenytoin (target voltage-gated sodium channels) and benzodiazepines (GABA-A receptor). Apparently, these drugs can modify the 'excitation-inhibition balance' in the neuronal network to such an extent that the seizure likelihood is decreased. The remaining group suffers from pharmacoresistant epilepsy and we are challenged to provide these patients with alternative treatments, for instance neurostimulation or surgery. There is also a large group of patients, where treatment with medication is initially satisfactory, but in the course of several years of treatment, treatment becomes more and more difficult, as a manifestation of pharmacoresistance (secondary pharmacoresistance). The mechanisms underlying pharmacoresistance are varied, including changes in synaptic efficacy, changes in pharmacodynamics or neuronal sprouting.

Although in most patients seizures stop within a few minutes by the action of various intrinsic inhibitory mechanisms, this is not invariably so. If seizures last longer than 5 min, the patient is in a *status epilepticus* (SE) [47]. During status epilepticus, seizures repeat with variable intervals or continue for long periods [123, 136]. In particular in patients with a generalized status epilepticus with tonic-clonic seizures, a status is a very serious medical condition. As during a generalized tonic-clonic status epilepticus, the respiratory muscles are involved, too, ventilation is limited causing cerebral hypoxia with risk of persistent neuronal damage. Treatment with medication (often Gaba-ergic medication, for instance Midazolam) is administered as a first line of treatment to stop the seizures. If this is not successful, patients are often treated in the ICU: as more medication is needed, the respiratory center is suppressed, as well, and mechanical ventilation is often needed.

During status, several changes in synaptic transmission occur. This is believed to be one of the mechanism involved in making it more and more difficult to treat the seizures as the status persists. Already within minutes after the seizure starts, *receptor trafficking* occurs: various ligand gated channels move from the synaptic membrane into endosomes, where they are inactive [79], as illustrated in Fig. 9.1. This process strongly limits the maximum channel conductance, g_{max}. As the GABA_A receptor is often involved in this process, with a time scale of minutes to hours, this is one of the key contributions to the refractoriness of a longer standing status epilepticus. Therefore, if a patient is in a status epilepticus, the sooner the intervention takes places, the more easy it is to stop the seizure.

9.3 Pathophysiology of Seizures

What are the mechanisms that change the physiological brain state to the pathological seizure state? And if seizures have started, why do they stop? These questions are clinically very relevant. For instance, if one could define a particular control parameter that is responsible for the transition to seizures, a candidate mechanism for control is available. In this case, seizures may result from bifurcations. It is also possible, however, that in some patients the brain is bistable: the brain can 'switch' between the normal and seizure state. Indeed, different routes to seizures exist. Lopes da Silva et al. discussed Model I-III transitions [43]. In Model I epileptic brains, random fluctuations of a particular variable are sufficient to start or stop seizures; such brains show bistable behaviour. An example is absence epilepsy. In Model II epilepsies, bifurca-

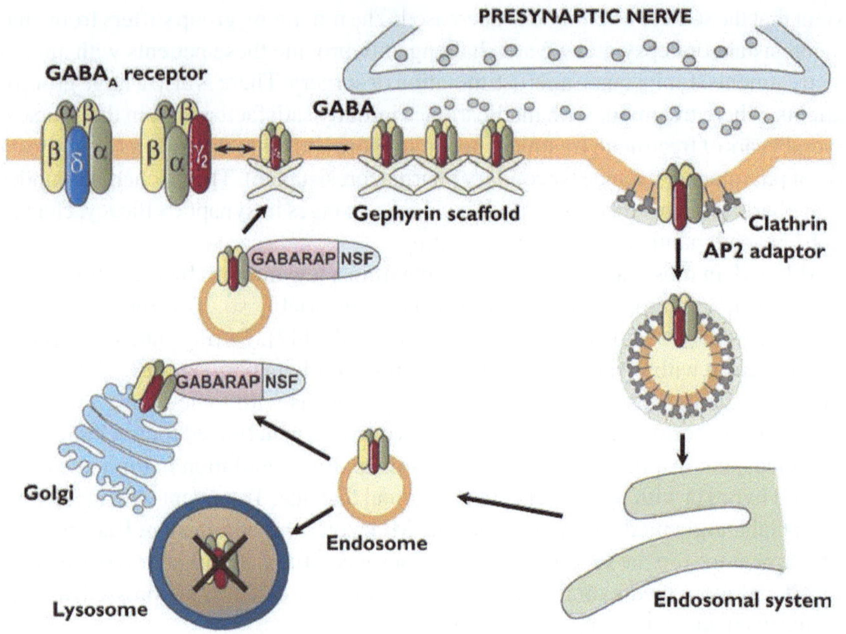

Fig. 9.1 Cartoon summarizing the role of receptor trafficking. After repeated seizures and massive γ-aminobutyric acid release, the synaptic membrane of GABA$_A$ receptors internalizes, inactivating the receptors. From: Chen and Wasterlain, Mechanistic and pharmacologic aspects of status epilepticus and its treatment with new antiepileptic drugs, Epilepsia, 2008. Reprinted with permission from John Wiley and Sons

tions occur as a particular endogenous variable slowly changes. Candidates are fluctuations in neurotransmitter concentrations or hormones. Model III epilepsies are characterized by induction of seizure by a defined external stimulus (and may include an endogenous variable, too): reflex epilepsies, for instance photosensitive epilepsy. We discussed the first two scenarios (Model I and Model II) earlier in Chaps. 3 and 4, here summarized as a cartoon in Fig. 9.2. We know a few epilepsy syndromes where seizures can be provoked by a particular external stimulus, for instance photic stimulation. In these patients periodic stimulation with (bright) light can induce generalized seizures: photosensitive epilepsies. This most likely resembles a bistable condition. In very young children, a fast rise in temperature may result in seizures.

9.3.1 Seizures Beget Seizures

In particular in the young brain, persistent focal seizure activity may induce new 'seizure foci' in other parts of the brain. Fundamental experiments were performed by

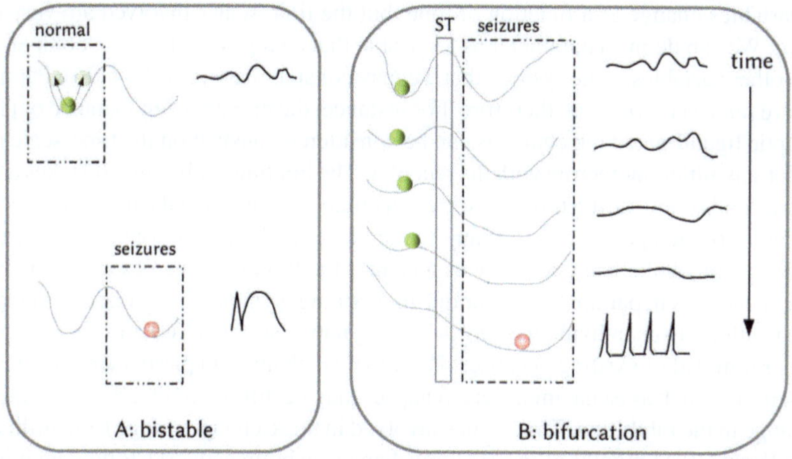

Fig. 9.2 Two possible scenarios involved in seizures or status epilepticus. Left: scenario A. The system has two equilibria. One reflects physiological function, while the other is associated with seizures. The system may switch between the two states, e.g., resulting from noisy 'endogenous' input (for instance synaptic noise), while the landscape remains unchanged. Right: scenario B. There is a qualitative change in the landscape towards another equilibrium state. The arrow of time has variable duration, from hours to days. The transition towards a very different dynamic behavior (seizures) may be caused by moving to another stable state (left) or may result from a change in control parameters (the shape of the landscape has changed): a bifurcation (right). ST: seizure threshold. Slightly adapted from [134]. Reprinted with permission from Elsevier Inc.

Ben-Ari, showing that kainate-induced 'seizures' in hippocampal brain slices were able to transform remote networks into an epileptogenic focus [66]. The induction mechanism required activation of excitatory (NMDA) receptors and long-term alterations in GABA-ergic synapses. These synapses were found to become *excitatory* because of a shift in the chloride Nernst potential, resulting from a change in the regulation of the chloride transmembrane ion gradients. Recall, that the inhibitory or excitatory characteristics of a ligand-gated receptor are primarily defined by the ion gradients of the ion conducted by the particular channel.

9.4 Models for Seizures and Epilepsy

From a dynamical systems point of view, we can view the brain as a high-dimensional dynamical system, defined by an independent set of *system variables*,[1] which evolve in time following a set of deterministic equations and *system parameters*, where the latter is a variable that is relatively constant. The distinction between a variable and a parameter in a biological system is not trivial. In fact, in a biological system nearly

[1] A variable can be defined as anything that can be measured, for instance the membrane potential

all variables change as a function of time, but the time-scales involved are very different. We can define a parameter as a variable that changes so slowly in comparison with other variables in the system, that we may consider it as a constant [76]. Parameters are *relatively constant*, therefore. For instance, the maximal conductance of post-synaptic ligand-gated ion channels can be considered constant on the time-scale of a neurotransmitter interacting with the receptor. The binding itself typically takes a few nanoseconds only; during this period we can reliably assume that the number of receptors will not change. However, many parameters do change on longer time scales, either in a physiological context or as a result of pathology. As an example, physiological changes in parameters occur during learning: synaptic transmission changes, both resulting from individual changes in the synaptic strengths as from generation of new or removal of existing synapses. Pathological changes in parameters may result from injury, such as ischaemia, were synapses may be differentially affected, causing a change in the inhibition. Time scales involved in these changes range from milliseconds (learning) to months as in structural changes in brain networks following injury.

In acquired epilepsies, persistent changes in parameters occur, typically evolving with time scales of weeks to years. This may involve a reduction in receptor densities, changes in cell morphology and function resulting from axonal sprouting, or expression of ion transporters. This process is known as *epileptogenesis*. Short-term changes in a parameter, for instance temperature, may induce the transition to a seizure in susceptible brains. An example are febrile seizures: in young children a fast increase in temperature may cause generalized seizures. Here, temperature is the control parameter, and the *febrile seizure* may result from a bifurcation.

Dynamical models aim to describe how variables change as a function of time. In epilepsy and epileptogenesis, the goal of these models is to advance our understanding of epileptogenesis or the transition of normal brain functioning towards a seizure. Models can be divided into several categories, for instance by a division of biological detail and scale, as illustrated in Fig. 9.3. Cellular models aim to elucidate mechanisms at the level of individual neurons or a small number of neuronal populations, and include ion channel conductances, synaptic inputs and the micro-environment, that may include dynamics of concentration gradients and non-constant Nernst poten-

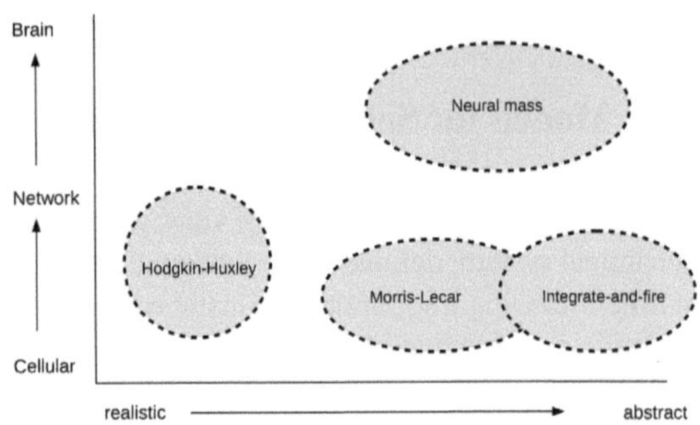

Fig. 9.3 Models can be categorized based on spatial scale and biological realism

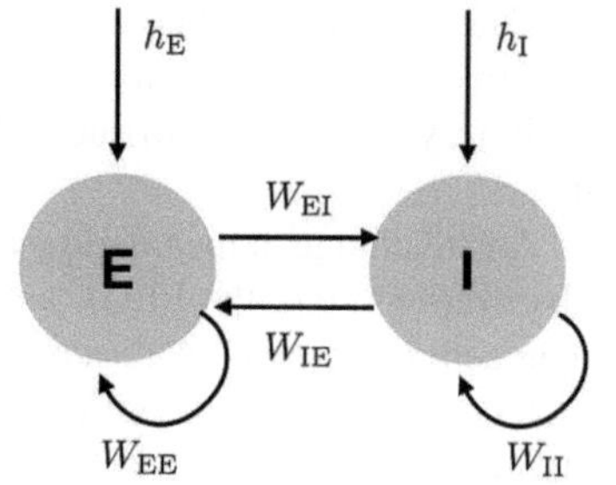

Fig. 9.4 The Wilson Cowan model consists of an excitatory (E) and an inhibitory (I) population, reciprocally coupled with weights W_{EI} and E_{IE}. Further, each population has an internal feedback loop, indicated with W_{EE} and W_{II} for the excitatory and inhibitory population, respectively. Each population receives external input, $h_i, i = E, I$

tials. An advantage of these models is that it can provide detailed understanding of molecular and cellular processes involved in epilepsy and seizures. It can also suggest novel therapeutic approaches that target particular ligand-gated or voltage-gated ion channels. A drawback, however, is that large scale phenomena, including the connection with scalp EEG recordings, or seizure spread, are not captured by such models, and that the number of variables in these models can be very large. Macro-scale models are concerned with dynamics of (large) sets of neuronal populations. This category comprises neural mass or meanfield models that we also discussed in Chap. 7. Excellent reviews about modeling neural activity, including applications for epilepsy are [39, 126].

The most elementary meanfield model consists of two sub-populations: one representing excitatory (E) neurons and one representation inhibitory (I) neurons, as originally proposed by Wilson and Cowan [142], and illustrated in Fig. 9.4. The dynamics was modeled to represent the proportion of neurons that were active at a particular moment of time, expressed as a two-dimensional system describing $E(t)$ and $I(t)$ for the excitatory and inhibitory population as

$$
\begin{aligned}
\tau_e \frac{dE}{dt} &= -E + \left(1 - r_e E\right) S_E\left[w_{EE}E - w_{EI}I + h_E\right], \\
\tau_i \frac{dI}{dt} &= -I + \left(1 - r_i I\right) S_I\left[w_{IE}E - w_{II}I + h_I\right],
\end{aligned}
\tag{9.1}
$$

with h_E and h_I are external inputs, τ_e and τ_i are time constants, r_e and r_i are constants describing the length of the refractory periods, S_x with with $x = E, I$ are sigmoid functions expressing the nonlinearity of the interactions, given by

$$
S_x = \frac{1}{1 + \exp(-b_x(u - \theta_x))} - \frac{1}{1 - \exp(b_x\theta_x)}. \tag{9.2}
$$

The synaptic weights, w_{EE}, w_{EI}, w_{IE}, w_{II}, represent the strength of the excitatory to excitatory, excitatory to inhibitory, inhibitory to excitatory, and inhibitory to inhibitory interactions. In one of the exercises you can explore the WC-model in more detail.

An advantage of meanfield models is that they can well connect with macroscopic observations (e.g. the EEG), and typically contain fewer variables and parameters than cellular models. A disadvantage is that detailed cellular characteristics are not included. This limits definition of potential treatment targets. A few papers, however, attempt to combine sub-cellular/cellular variables of detailed models and lumped parameters contained in meanfield models. For instance, a neural mass model based on single cell dynamics was recently proposed in [143]. With this approach, it was possible to explore the influence of single cell parameters on network activity, which is not possible in traditional neural mass models. It was for instance simulated that an increase in extracellular potassium resulted in a loss of equilibrium and occurrence of large amplitude limit cycle activity of the populations started. This increase in extracellular potassium has indeed been observed during microseizures at the onset of electrographic seizures [107].

9.5 Detailed Models for Seizures

The number of neurons contained in detailed models ranges from as few as two cells [101] to several hundreds [129], one million [60] or even more, as explored in the blue brain project.[2] The level of detail required in a model depends on e.g. the mechanism one wishes to study, the computational need, and the availability of biological detail to guide the modeling. For instance, one could study how a particular balance of inhibition and excitation on a small scale (2 neurons) generalizes to a larger population of cells, with *intrinsic differences between neurons*. Such heterogeneity is present in biological systems, and may include variations in conductances g or time constants. To this end, Skinner and coworkers [101] simulated a two-cell system to predict N-cell (N = 20, 50, 100) network dynamics for heterogeneous, inhibitory populations. They observed that key characteristics of the dynamics studied with the n = 2 model were preserved in larger networks. For instance, multistable patterns in the two-cell system corresponded to different and distinct coherent network patterns in the larger networks for the same parameter sets.

Modeling may also suggest new hypothesis by elucidating particular mechanisms not previously considered. As an example, it is generally assumed that seizures result from an increased excitation/inhibition ratio. A study by Van Drongelen et al. [128, 129] was at variance with this assumption. They modeled a neocortical network consisting of 656 neurons (512 pyramidal cells and 144 interneurons). This model showed seizure-like behavior when the synaptic excitation to both inhibitory and excitatory cells was *decreased* from baseline. Perhaps the reduced excitation resulted in an effec-

[2]See http://bluebrain.epfl.ch/.

tive reduced firing rate of interneurons and hence reduced relative inhibition in the network (feedforward inhibition). These findings are supported by various experimental and clinical observations. For instance, in patients with typical absence seizures, it is generally assumed that impaired γ-aminobutyric acid (GABA)-ergic inhibition is involved in the pathophysiology, effectively resulting in increased excitation. However, in diverse genetic and pharmacological models of absence seizures the extrasynaptic GABA-A receptordependent tonic inhibition was shown to be *increased* in thalamocortical neurons [26]. It is also known that several anti-epileptic drugs can *increase* the seizure likelihood in some epilepsy syndromes. For example, carbamazepine and phenytoin block sodium channels by binding preferentially to the inactivated state. In patients with absence seizures, however, they can exacerbate the seizure frequency. In a network model studying this phenomenon, it was shown that action potential frequencies increased after 'application' of voltage-gated sodium channel blockers [118]. These two examples illustrate that predicting drug efficacy in epilepsy is essentially impossible without considering the network properties. Indeed, the remarkable finding of the studies by Van Drongelen [128, 129] illustrate the value of computational modeling in epilepsy. As the number of nonlinear interactions between neurons is very large, intuition is generally inadequate to predict or define the relevance of particular variables in the generation seizures. Indeed, the intuitive notion that increased excitation alone would cause seizures is too simplistic for a network of interacting neurons with many feedback loops, including feed-forward inhibition. These non-trivial effects on overall network behavior, in part resulting from non-selective targeting of neurons, most likely also explains the pharmacoresistance in about 30% of patients with epilepsy.

9.6 A Meanfield Model for the Transition to Absence and Non-convulsive Seizures

Absence seizures are characterized by a sudden change in behaviour and awareness where the patient (typically infants, children or young adults) show brief periods of behavioural arrest. The characteristic EEG pattern shows bilateral synchronous (generalized) spike and wave discharges. The frequency is initially approximately 3 Hz, but always reduces to a lower frequency at the end of the absence period,[3] that typically lasts has 2–30 s. We showed such pattern in Fig. 3.1. Absence seizures are genetically determined and may for instance results from a change in the GABA-a receptor. Absence seizures result from abnormal interactions between the thalamus and various

[3]This is clinically relevant, as well: in some patients, EEGs can show rhythmic behavior where one can doubt if this is ictal activity. If frequencies do not change, it is most likely an artifact. A classic example is 5–7 Hz rhythmic activity that is sometimes observed in ambulatory recordings. This can result from movement of the head during brushing the teeth, and is known as the "toothbrush artifact".

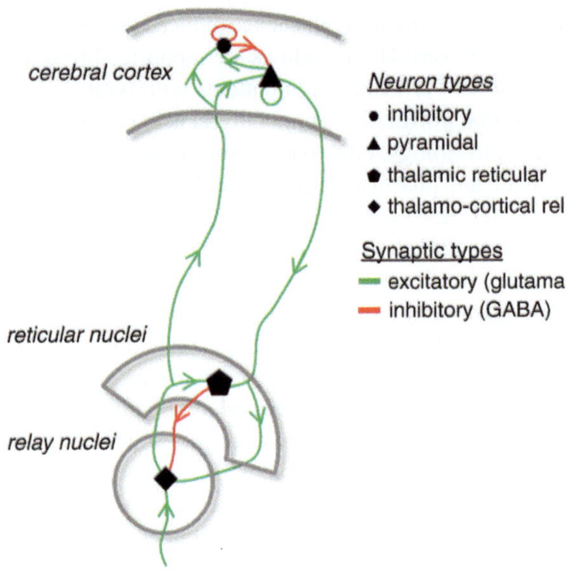

Fig. 9.5 Synaptic organization of the thalamocortical model. Illustrated are the synaptic pathways that connect the different neuronal populations within the model. The model contains four types of neuronal populations which are connected through excitatory (green) and inhibitory (red) synaptic projections. The inhibitory connection between the reticular and the relay nuclei in the thalamus is comprised of both a fast GABA-A and a slow GABA-B synaptic connection; the latter connection can be modeled with a time-delay. Similar, a time delay can also be included for the cortico-thalamic connections. Illustration from [51], Reprinted with permission from Elsevier Inc.

cortical areas and several essential components have been identified, as illustrated in Fig. 9.5 In patients during a non-convulsive status similar clinical and EEG phenomena may be observed. We describe a case from our own clinic.[4]

> On April 1, 2011, a 65-year old patient was seen at the emergency department of the hospital Medisch Spectrum Twente, Enschede, The Netherlands. History taking was hardly possible, due to a severe dyspnoea. Oxygen saturation was 88%, with an arterial $pO_2 = 7.3$ kPa, $pCO_2 = 6.8$ kPa and a pH = 7.50. He was initially diagnosed with heart failure and treated with diuretics. His condition hardly improved, however, and he remained confused. Soon after, he developed a pneumonia with a respiratory insufficiency, for which he was transferred to the Intensive Care Unit on April 3, 2011. He was intubated and mechanical ventilation was started. He was sedated with continuous infusion of propofol. After a few days, his respiratory condition gradually improved, and sedation was stopped. A few hours later, he suffered from a generalized tonic-clonic seizure, for which the neurologist was consulted. On clinical examination, his Glasgow coma scale score was minimal. There were no abnormal eye movements, and pupil size was normal, with intact reactions to light. A brain CT showed moderate generalized atrophia and diffuse white matter abnormalities, without any signs of recent ischaemia or hemorrhage. Under suspicion of a possible encephalitis, a

[4]Text is literally from our paper in [51]. Frontiers is fully compliant with open access mandates, by publishing its articles under the Creative Commons Attribution licence (CC-BY).

lumbar puncture was performed. The opening pressure was normal, and the cerebrospinal fluid revealed no significant abnormalities.

The patient was treated with diphantoine, but consciousness did not return. The differential diagnosis included a non-convulsive status epilepticus (NCSE) and continuous EEG recording was started. This showed rhythmic, high voltage (150 μV) delta activity, with a left hemispheric dominance. Sometimes, spikes were observed, as well. This pattern was interpreted as electroencephalographic seizure activity. After about 40 min, the rhythmic delta activity evolved into rhythmic spikes, and the patient suffered from a second generalized seizure. Propofol was restarted, but non-convulsive seizure activity continued. Therefore, midazolam and valproic acid were added, too. Eventually, after 2 days, all epileptiform discharges disappeared. After gradual reduction of the sedation with propofol and midazolam, our patient eventually recovered consciousness. Initially, he showed a severe bradyphrenia, with a mild right- sided hemiparesis. A repeat CT cerebrum showed two subcortical infarcts in the right hemisphere, that could not explain his mild right-sided paresis. After a week, he was successfully weaned from the ventilator and his condition further improved. He was discharged from our ICU on April 14, 2012. In sum, this 65-year old patient suffered from both convulsive and non-convulsive seizure activity, where the EEG showed rhythmic delta activity with intermittent spikes. This is a relatively rare EEG pattern, that should not be interpreted as post-ictal slowing, but as an ictal phenomenon. The corresponding spectrogram is shown in Fig. 9.6a. The time-series were filtered between 0.5 and 20 Hz using a fourth-order zero-phase Butterworth filter. For subsequent analysis, we selected the high signal-to-noise epochs marked by white circles; epochs are shown in Fig. 9.6b.

Both in patients with absence seizures and in our patient these rhythmic discharges can be modeled using the thalamocortical model shown in Fig. 9.5. The model comprises four types of neural populations: cortical pyramidal cells, cortical inhibitory neurons, thalamo-cortical relay cells and thalamic reticular neurons. This model describes the dynamics of the average membrane potentials V_k and firing rates Q_k of the four different populations and can produce periodic patterns similar to clinical recordings. A comparison between an actual recording and the EEG generated by the TC model is shown in Fig. 9.7. Several other implementations of the thalamocortical model exist, where variations include the presence or absence of time delays within the thalamus (GABA-B connections between reticular and relay nuclei) or the cortico-thalamic connections. For instance, in [74], the critical dependency of the onset of the spike-wave activity on (i) the coupling strength ν_{se} between cortical pyramidal cells and thalamic relay cells and (ii) the time delay τ from the reticular to the relay neurons, mediated by the slow GABA-B synapses, was studied. These two parameters were shown to be critically involved in the generation of epileptiform discharges, illustrated in Fig. 9.8. Most absence seizures tend to disappear as children grow older; we could speculate that this is accompanied by changes in these parameters. A few examples of resulting time series for different combinations of the time delay for a fixed value of ν_{se} are shown in Fig. 9.9.

Spike-wave discharges can also be modeled taking the approach presented in [116]. In this work, authors introduce a meanfield thalamocortical model comprised of pyramidal cells, interneurons, thalamocortical cells and reticular cells. The model displays *bistability*, similar to a previous model by [110], and transitions between the physiological state and the generation of spike-wave discharges are modeled by stochastic events. These two approaches to model spike-wave discharges are fundamentally

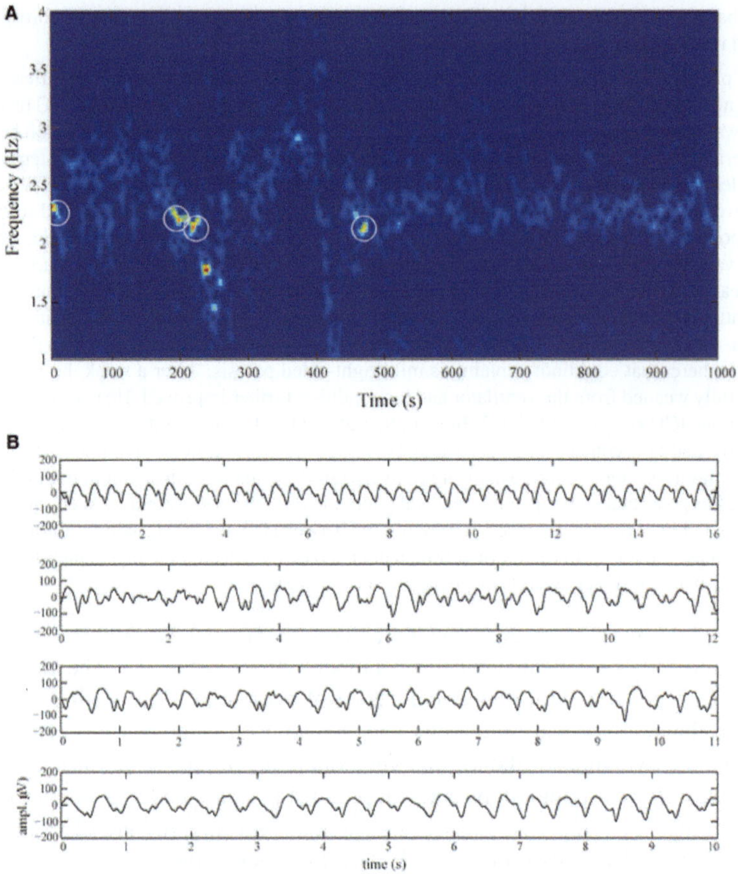

Fig. 9.6 Diffuse rhythmic delta activity during non-convulsive status epilepticus. **a** Spectrogram of electrode Fz over the entire recording-length. The activity is almost completely confined to frequencies in the delta band (1–3 Hz). The four white circles schematically denote the selected epochs. These epochs are shown in (**b**), displaying rhythmic delta activity with frequencies between 2 and 2.5 Hz and amplitudes up to 150 μV. The oscillations are dispersed with intermittent spikes. Illustration from [51]

different, as in the model introduced in [74] bifurcations occur between physiological activity and spike-wave discharges, while [110] and [116] introduce a bistable system, and transitions result from stochastic events. This difference is clinically relevant, too. If the seizures result from a bistable system, prediction of seizures is essentially impossible. A change in a particular parameter that results in a bifurcation can (at least theoretically) be recorded, with potential for seizure prediction. Students are encouraged to read the original papers and study the bifurcation diagrams in the original papers. Simulations of this bistable model are shown in Fig. 9.10. Recently, another mechanism was proposed for spontaneous transitions between ictal and inter-

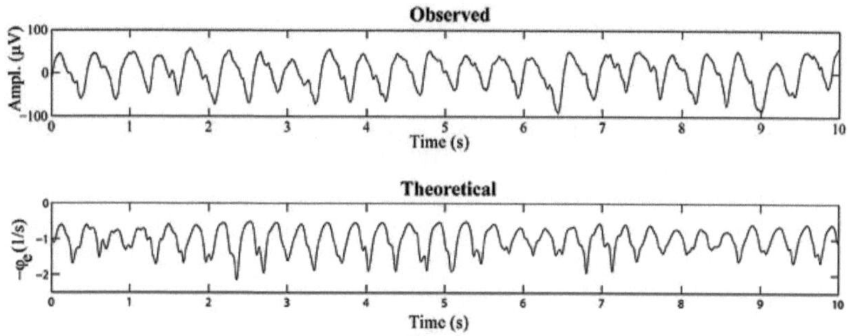

Fig. 9.7 10-s epochs of recorded (top-row) and simulated (bottom-row) EEG time-series. There is a satisfactory agreement between the observed time series and the simulations. Illustration from [51]

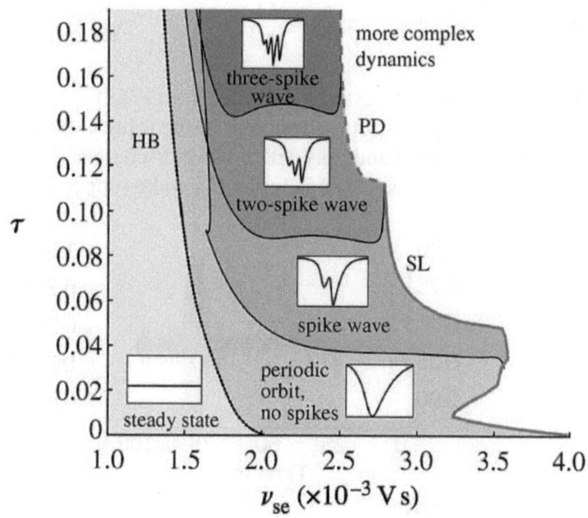

Fig. 9.8 Two parameter continuation in the $\nu_{se} - \tau$-plane. Transitions from the steady state to the periodic orbits are mediated by Hopf bifurcations (HB). Spikes can be present in the orbits, that can become unstable via periodic doubling (PD) or disappear via a saddle node bifurcation (SL). Illustration from [75]. Reprinted with permission from the Royal Society

ictal activity was proposed by introducing activity dependent synaptic depression and recovery, in combination with a particular network architecture [62]. Using a 2D single-layer neural network model comprised of 10,000 neurons (9600 pyramidal neurons, interspersed with 400 interneurons) it was shown that simply as a consequence of ongoing activity at synapses that undergo activity-dependent depression and recovery, such transitions are indeed possible.

While the various models introduce candidate mechanisms for seizures, the clinical impact of these models is still limited. However, this may change in the next decade as models improve, and may be tuned to individual characteristics for instance to assist in optimal choice of medication.

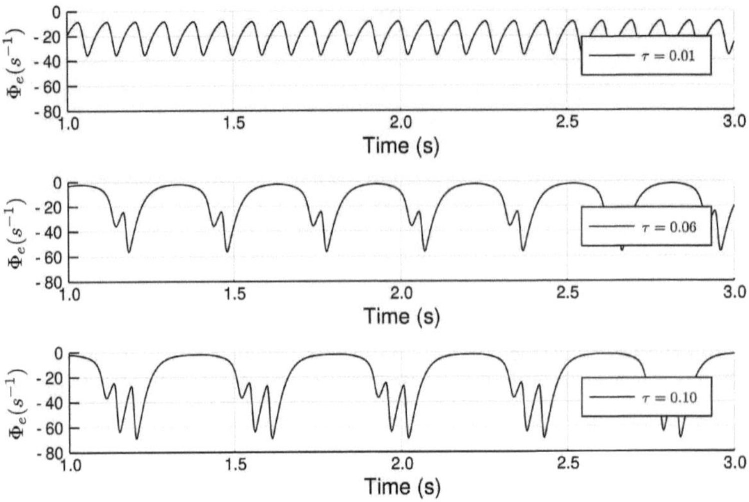

Fig. 9.9 Simulations using the model from [74] with $\nu_{se} = 0.0025$ Vs and three different values of the time delay τ from the reticular to the relay neurons. If the delay is sufficiently large, spike-wave (middle panel) and poly-spike-wave discharges (lower panel) occur. For $\tau = 0.01$, the model generates periodic oscillations without a spike (upper panel)

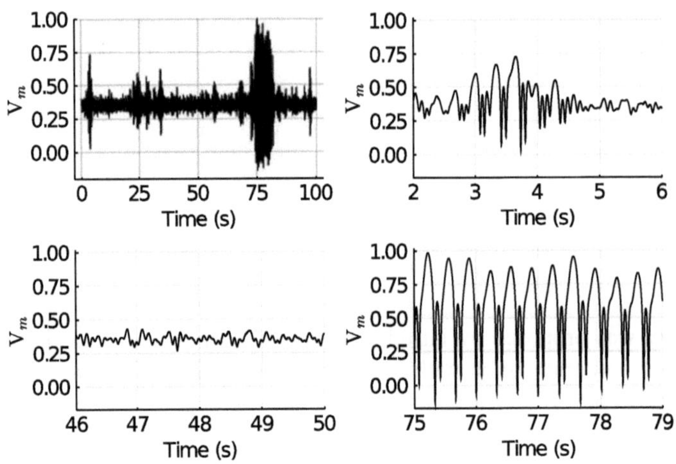

Fig. 9.10 Illustration of simulations of EEG rhythms (upper panel) using the model from [116]. Spike-wave discharges occur if the system jumps from the 'physiological state' (left lower panel) towards the other stable state, where the transitions result from stochastic events. This is most pronounced in the interval between 75 and 85 s (right lower panel).The system also generates polyspike waves (right upper panel). The model was implemented in the Julia programming language (https://julialang.org/)

9.7 Treatments for Epilepsy

The first line of treatment for patients with epilepsy is with medication. Over the last century, a large number of anti-epileptic drugs (AEDs) have been developed, with varying mechanisms of action. In general, at the excitatory synapse, AEDs can interact with voltage-gated Na^+-channels, synaptic vesicle glycoprotein 2A (SV2A), voltage-gated Ca^{2+}-channels, AMPA and NMDA receptors. AEDs can also exert effects at inhibitory synapses. This includes a direct interaction with the GABA receptor at the postsynaptic density or inhibition of the GABA transporter, leading to a decrease in GABA uptake into presynaptic terminals and surrounding glia. This is summarized in Fig. 9.11. Unfortunately, adequate treatment with drugs is only effective in approximately 70% of the patients with epilepsy, a percentage that has essentially not changed over the last 30–50 years. This unchanged percentage of pharmacoresistance most likely results from insufficient selectivity of anti-epileptic drugs within the particular neural network that is involved in the generation of seizures. Drugs that are beneficial in some epilepsy syndromes may even result in an increase in seizures in others [118]. This motivates alternative treatments, including deep brain and vagus nerve stimulation and epilepsy surgery. We will discuss some aspects of vagus nerve stimulation for epilepsy in the next chapter.

9.7.1 Epilepsy Surgery

The goal of epilepsy surgery in patients with pharmacoresistant epilepsy is to remove the 'epileptogenic zone'(EZ). The EZ is defined as the smallest area of cortex that should be removed to obtain seizure freedom. In practice, it is approximated by the 'seizure onset zone' (SOZ): that part of the cortex that shows interictal spikes, and is the presumed origin or generator of the seizure.

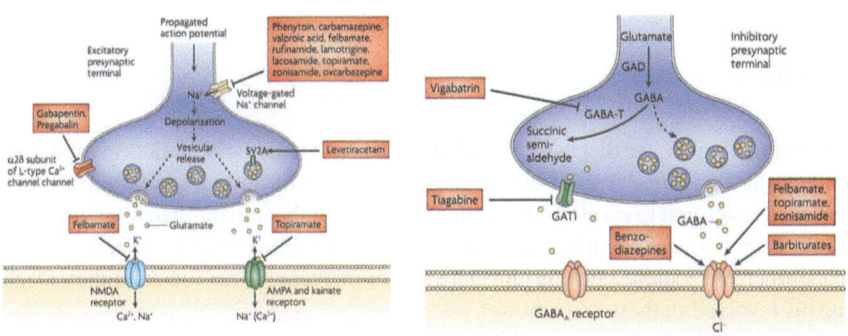

Fig. 9.11 Left: targets for anti-epileptic drugs at the excitatory synapse. Right: targets for anti-epileptic drugs at the inhibitory synapse. Reprinted from [13], with permission from Springer Nature

The underlying assumption that the "generator" of a seizure is limited in space may not always hold true. In fact, it is not uncommon that the presumed SOZ is removed, but the patient still has seizures. In recent work, a computational model was created that consisted of a neuronal mass that modeled the SOZ and three other masses representing other, connected brain areas. In the simulations it was shown that for particular choices of the model parameters, removing the SOZ did not always result in 'seizure freedom'. In fact, removal of normal populations located at a crucial spot in the network was typically more effective in reducing seizure rate. Network structure and connections may be more important than localizing the SOZ, and this may explain why removing of the SOZ is not always effective [50].

9.7.2 *Assessment of Treatment Effects*

Another challenge in the treatment of patients with epilepsy is how to assess the effect of the therapy as a reliable bio-marker does not exist.[5] Indeed, evaluation of the treatment response mainly depends on reported seizures by the patient (or its caregiver). This has been shown to be quite unreliable [48]. Exciting new developments, using small subcutaneous electrodes that allow 24/7 recording of EEG activity, may allow objective seizure counting [140]. This can benefit optimisation of treatment with anti-epileptic drugs. Further, these ultra-long EEG recordings may contain information for seizure prediction in some patients.

9.8 Summary

We discussed seizures and epilepsy and how biophysical and mathematical models may assist in furthering our understanding of seizures and the transition to seizures.

Problems

9.1 In this chapter, we discussed detailed models and meanfield models in relation to epilepsy and seizures. A third model category is an 'educational model'. This question challenges you to create an educational model for the transition to a seizure using the the Morris-Lecar equations. Can you simulate bistability in a ML model? And the transition from a stable equilibrium state to a limit cycle? What are candidate mechanisms involved in these transitions?

[5] For treatment of high cholesterol, one can simply assess the cholesterol concentration in the blood. While serum concentrations of anti-epileptic drugs can be measured, this correlates poorly with treatment efficacy.

Take $I = 0.075$, $\mathrm{C} = 1$, $E_k = 0.7$, $E_L = 0.5$, $E_{\mathrm{Ca}} = 1$, $g_{\mathrm{k}} = 2$, $g_{\mathrm{k=Ca}} = 1$, $g_{\mathrm{L}} = 0.5$, $V_1 = 0.01$, $V_2 = 0.15$, $V_3 = 0.1$, $V_4 = 0.145$ and $\phi = 1.15$; for units, see Table 4.2. You should find a stable fixed point, a saddle point and an unstable fixed point.

Show that a stable limit cycle exists, and that if you decrease the current I the periodic orbit grows in amplitude and comes closer to a saddle point. The period increases, as well, and near the homoclinic bifurcation, where the orbit collides with the saddle at $I = I_c$, the frequency of oscillation scales as $1/\log(I - I_c)$. For details, see also [4], but note there is a mistake in some of the parameter values given in the caption of Fig. 2.4 in Sect. 2.3 of this paper (communicated with the authors; the correct values are presented in this Exercise).

9.2 How can a change in extracellular potassium result in seizures? Study the paper by [143] and references therein to suggest a model.

9.3 Consider Izhikevich's reduction of the Hodgkin-Huxley equations that we discussed in Chap. 4, given by

$$\dot{v} = 0.04v^2 + 5v + 140 - u + I$$
$$\dot{u} = a(bv - u)$$

with the additional condition that if $v = 30$ mV then $v = c$ and $u = u + d$. Take $a = 1, b = 1.5\ c = -60, d = 0$ and $I = -65$. Show that the system has two equilibria: a stable equilibrium and a limit cycle. Show that you can perturb the system with an additional inhibitory current to prevent limit cycle oscillations. Hint: visit https://www.izhikevich.org/publications/whichmod.htm#izhikevich

9.4 Febrile seizures result from a fast increase in temperature, typically up to 40 °C. Can you suggest potential mechanisms why a temperature change could result in changes in neuronal excitability, resulting in seizures?

9.5 A potential mechanism involved in the generation of a mirror focus in epilepsy is discussed in [66]. Can you simulate this process in a simple model?

9.6 Use the Matlab code WC.m to simulate how the Wilson-Cowan equations can show oscillatory behaviour of both the excitatory and inhibitory neuronal populations.

9.7 Photosensitive epilepsy is a generalized epilepsy where seizures can be induced by stroboscopic light flashes in the frequency range 5–25 Hz. Can you make a simple model to simulate seizure induction by such an external stimulus?

Take $I = 0.075$, $C = 1$, $E_L = 0.7$, $E_K = 0.5$, [illegible] $= 1$, [illegible] $= 0.5$, [illegible] $= 0.01$, [illegible], and $\phi = 1.15$, for units see Table 4.2. You should find a stable fixed point, a saddle point, and an unstable fixed point.

Show that a stable limit cycle exists and that if we decrease the current I the periodic orbit grows in amplitude and comes closer to a saddle point. The period increases, [illegible] with the saddle at [illegible]

[illegible]

Part VI
Neurostimulation

Chapter 10
Neurostimulation

Vulnerability is the least celebrated emotion in our society

— Mohadesa Najumi

Abstract Brains function at a delicate balance between excitation and inhibition. Changes therein may result in pathology, such as seizures, chronic pain and Parkinson's disease. We have increasing possibilities to modulate this balance by electrical stimulation, although most working mechanisms are poorly understood.

10.1 Introduction

Neurostimulation can be defined as modulating particular functions of the nervous system using electrical currents. Neurostimulation is currently used for the treatment of various neurological and psychiatric disorders. Examples include Parkinson's disease, epilepsy, migraine, neuropathies and major depressive disorder. Stimulation of the central nervous system is also used for diagnostic purposes. An example is peri- and intra-operative stimulation of the cortex to delineate the epileptogenic zone or assessment of cortical excitability with transcranial magnetic stimulation in patients with (presumed) epilepsy.

10.2 Neurostimulation for Epilepsy

In approximately 30% of patients, seizures cannot be adequately controlled with medication. In a fraction of these patients, neurostimulation can reduce the seizure frequency [78]. In noninvasive stimulation, currents are applied to the intact skin. Low intensity stimulation, using transcranial electric stimulation (TES), has been

M. J. A. M. van Putten, *Dynamics of Neural Networks*,
https://doi.org/10.1007/978-3-662-61184-5_10

shown to alter excitability of cortical neurons. In particular, transcranial direct current stimulation (tDCS), using transcranial currents of the order of a few mA, has been shown to modify seizure frequency in animal studies and *in vitro*. Another noninvasive technique that is being explored in epilepsy is transcranial magnetic stimulation (TMS), both for diagnostic and therapeutic purposes [32]. At present, however, both techniques have not been extensively evaluated in clinical trials. A recent Cochrane review discussing repetitive TMS in epilepsy reported that the technique is safe, but efficacy for seizure reduction is essentially lacking [23]. In contrast to noninvasive neurostimulation, invasive neurostimulation has shown to be effective in about 30% of the patients with pharmacoresistant epilepsy. The two most common techniques used are vagus nerve stimulation and deep brain stimulation (DBS).

10.2.1 Vagus Nerve Stimulation

Treatment of epilepsy with vagus nerve stimulation (VNS) dates back to 1880. At that time, DL Corning believed that changes in the cerebral blood flow caused seizures which could be modulated with vagus nerve stimulation. A century later, in 1988, the first chronic implantable stimulator was used to treat drug-resistant epilepsy. The FDA approved this technique in 1997 to treat partial onset seizures in pharmacoresistant patients. The left vagus nerve is stimulated by electrodes that are wrapped around the nerve; the pulse stimulator is placed just below the clavicle. The stimulator typically periodically stimulates the vagus nerve with currents in the range 1.0–3.0 mA, frequencies of 20–30 Hz and pulse widths of 130–500 ms. The reason for left vagus nerve stimulation is the innervation by the sinoatrial node by the right vagus nerve, thus reducing undesired cardiac-side effects. The most common side effects are dysphonia, hoarseness, and cough, resulting from stimulation of efferent fibers contained in the vagus nerve to the vocal cords. In some patients, these side-effects can be reduced by changes in the stimulation settings. Response rates between patients vary, but approximately 40% of the patients report a seizure reduction of 50% or more after 2–3 years of treatment. As of 2015, more than 100.0000 VNS devices have been implanted in patients with epilepsy.

10.2.2 Deep Brain Stimulation in Epilepsy

Deep brain stimulation (DBS) for epilepsy is often targeted towards the anterior nucleus of the thalamus [46] but many other stimulation sites, including amygdala, hippocampus and cerebellum, are currently explored in clinical studies [78].

10.2.3 Working Mechanism

It is still unclear how VNS and DBS result in a reduction of seizures. Candidate mechanisms in VNS include an increased activity in the locus coeruleus and raphe nuclei and release of norepinephrine and serotonin, which are known to have anti-epileptic effects. However, this explanation is not really satisfactory as it does not elucidate how this on its turn affects underlying dynamics. Similar arguments apply to reported effects of VNS on the limbic system, thalamus and thalamocortical projections, that are all involved in modulating 'cortical excitability'. Clearly, there is ample room to study this in more detail, which may also contribute to a pre-clinical identification of responders and non-responders.

DBS may disrupt abnormal hypersynchrony of neural networks in a frequency dependent manner. Electrical stimulation also releases neurotransmitters, that may change the excitation/inhibition ratio. It is also important to differentiate between acute effects, as relevant in closed-loop responsive neurostimulation where stimulation is a function of an impeding or actual seizure, and long-term effects.

To advance the use of neurostimulation, there is a clear need for further understanding of epilepsy and seizures. Computational modeling, experimental studies in animals and clinical trials can all contribute. Clinical challenges include the definition of the target for the stimulation, stimulation settings (frequency and current), and the choice between responsive or non-responsive neurostimulation. As currently approximately 30% of the patients experience a significant reduction of seizure frequency (more than 50%) when treated with neurostimulation, pre-operative identification of responders and non-responders is another challenge.

10.3 Neurostimulation for Parkinson's Disease

Parkinson's disease (PD) is a neurodegenerative disorder, resulting from a reduction in dopamine production from neurons in the substantia nigra. The dopamine deficiency results in a variety of clinical signs and symptoms, including tremor, rigidity, bradykinesia (slowing of movement) and postural instability. Also, non-motor symptoms occur due to PD including cognitive dysfunction and language problems. The first line of (symptomatic) treatment is administration of a dopamine precursor, Levodopa (combined with carbidopa, a peripheral decarboxylase inhibitor) that crosses the blood-brain-barrier and is converted to dopamine. After several years of treatment, however, effects may decline. Side effects are known, as well, including hyperkinesia. As an alternative, deep brain stimulation may be beneficial. Neurostimulation for Parkinson's disease (PD) dates back to the its incidental discovery by Alim Louis Benabid and Pierre Bolak in 1987. During a routine radiofrequency lesion (thalamotomy) of the nucleus ventralis intermedius of the thalamus to treat a patient with essential tremor they discovered that electrical high-frequency ($>$100 Hz) stopped the tremor. Stimulating with frequencies between 30 and 50 Hz did not have a significant effect.

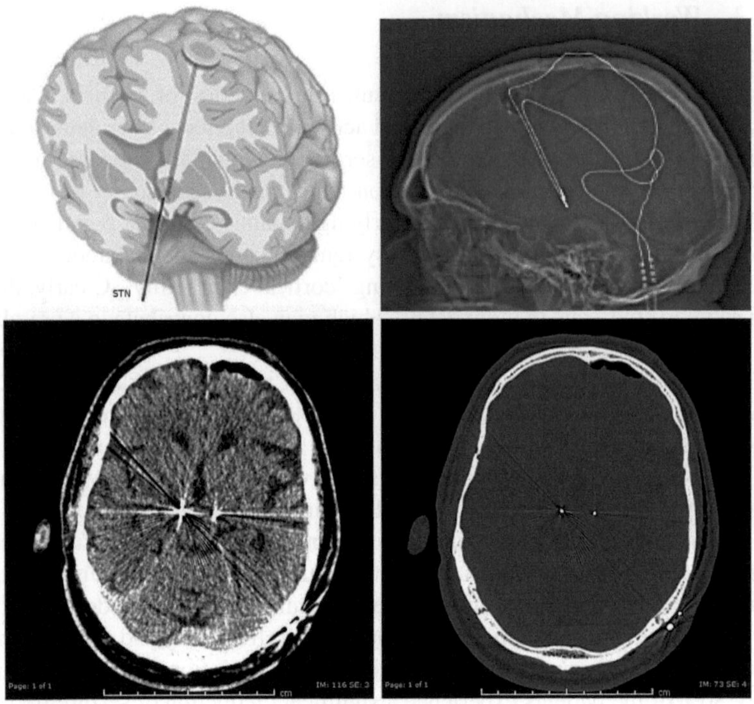

Fig. 10.1 Top left: Illustration of the placement of the electrode in the STN for the treatment of Parkinson's disease. Reprinted with permission from: *Systems approaches to optimizing deep brain stimulation therapies in Parkinson's disease*, Santaniello, Gale, and Sarma, Wires Systems Biology and Medicine, 2018. Top right: topogram of a patient who received bilateral DBS. The electrode tips are near the s. nigra. Bottom left: CT-cerebrum, showing electrodes. Note the artifacts from the electrode material. Bottom right: same CT, but now in bone setting. How can you deduce that the CT scan was made shortly after the surgery? Courtesy of Dr M. Tjepkema-Cloostermans, technical physician, and M. Hazewinkel, radiologist, Medisch Spectrum Twente, Enschede, the Netherlands

Soon after, this was successfully reproduced other patients. This discovery resulted in a very successful development of DBS in patients with PD and other movement disorders. At present, the main surgical targets are the subthalamic nucleus (STN) (Fig. 10.1) and the globus pallidus internal (GPi). Recently, the pedunculopontine nucleus (PPN) has been explored, as well. Stimulation of the PPN appears to be relatively beneficial for the treatment of primary gait and posture symptoms in PD. Despite the success, the working mechanisms are still poorly understood.

10.4 Spinal Cord Stimulation for Neuropathic Pain

Spinal cord stimulation (SCS) is a proven effective therapy for various types of mixed neuropathic conditions. A very common cause of a painful neuropathy is diabetes. Medication has often limited effects, and side effects are common, for instance drowsiness. A recent multicentre trial has shown that SCS is an effective treatment in patients with a painful diabetic neuropathy. At baseline, the average pain score, assessed with the visual analogue scale (VAS) was 73 in the SCS group and 67 in the control group. After 6 months of treatment, the average VAS score was significantly reduced to 31 in the SCS group ($P < 0.001$) and remained 67 ($P = 0.97$) in the control group. Patients who received SCS also reported improved health and quality of life after 6 months of treatment [33, 34].

10.5 Neurostimulation for Psychiatric Disorders

Several neuromodulatory techniques for the treatment of depression are currently available, including electroconvulsive therapy (ECT), vagus nerve stimulation (VNS), transcranial magnetic stimulation (TMS) and deep brain stimulation.

10.5.1 Electroconvulsive Therapy for Major Depression

ECT is the oldest neurostimulation therapy for the treatment of depression, and dates back to 1938. Response rates are very good, up to 60%. During the treatment period, generalized (tonic-clonic) seizures are induced by electrical stimulation while the patient is receiving generalized anesthesia and is paralyzed. The seizure typically lasts 1–3 min, and terminates by itself. During the procedure, the EEG is registered using two frontal (Fp1 and Fp2) electrodes. An example is shown in Fig. 10.2. The treatment consists of 15–20 sessions, for 1–2 times/week. Despite its long and established position in the treatment of severe depression, the working mechanism is still unclear. Side effects include memory disturbances, in particular short term memory loss.

10.6 Neurostimulation for Diagnostic Purposes

Neurostimulation is also used for diagnostic purposes, for instance to study 'cortical excitability' in patients with epilepsy [32] or the effects of drugs on cortical excitability [30, 88, 89]. Cortical excitability can be defined as the strength of a defined response of cortical neurons following a defined stimulus. A non-invasive

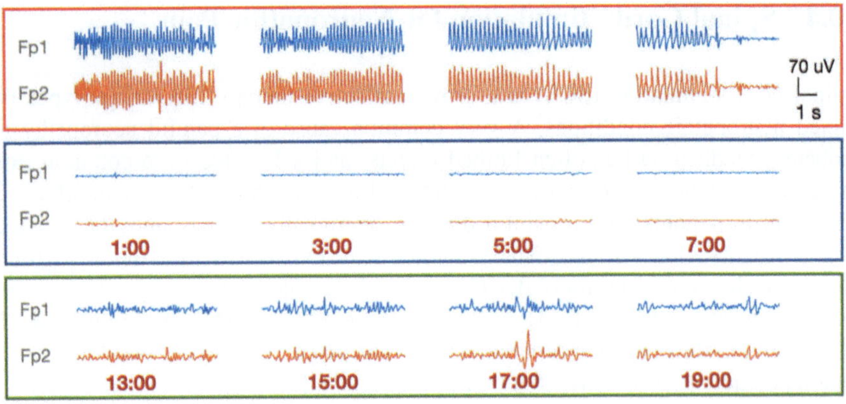

Fig. 10.2 EEG during ECT. Top row: ECT-induced seizure activity. Middle row: postictal suppression. Bottom row: recurrence of EEG rhythms. The times (min) indicated are relative to the end of the seizure. Courtesy of Dr J. van Waarde, psychiatrist, Rijnstate Hospital, Arnhem, The Netherlands

technique is transcranial magnetic stimulation (TMS), where a short duration magnetic pulse is created by a strong current pulse passed through a coil positioned above the head. If the change in magnetic field strength is sufficiently large, the generated electric field[1] is strong enough to induce transmembrane currents that can depolarize cortical neurons in a relatively painless fashion. The response thus generated can then be recorded with a scalp EEG (TMS-EEG). The induced EEG response, a transcranial evoked potential (TEP), can obtained after averaging sufficient responses, illustrated in Fig. 10.3). The different peaks and troughs in the TEP also change in response to particular drugs [30]. For example, the amplitude of the P180 is decreased by voltage-gated sodium channel blockers (such as lamotrigine or carbamazepine) and a decrease in the N100 can result in response to GABA-ergic drugs like diazepam. Alternatively, one can record the motor responses (the motor evoked potential, MEP) from e.g. a hand muscle if the TMS stimulus is applied above the motor cortex: TMS-EMG. The resting motor threshold (rMT) is defined as the minimum TMS pulse intensity needed to elicit reproducible MEPs in a target muscle. By using paired-pulse TMS, and varying the interstimulus interval (ISI, range 5–300 ms), one may obtain response curves as a function of the ISI. Here, the first TMS pulse serves as the conditioning pulse. Very short ISI can result in facilitation: the motor response is increased. ISI in the range 30–250 ms can cause inhibition, reflected by a decrease in the motor response. These methods are summarized in Fig. 10.3. While many studies report changes in the resting motor threshold and shape of the TEP response in patients with epilepsy [7, 10, 117] and effects of drugs [88], not all findings appear to be robust [6, 9]. In addition, some expressed concerns about potential

[1] Recall the Maxwell-Faraday equation $\nabla \times \mathbf{E} = -\partial \mathbf{B}/\partial t$.

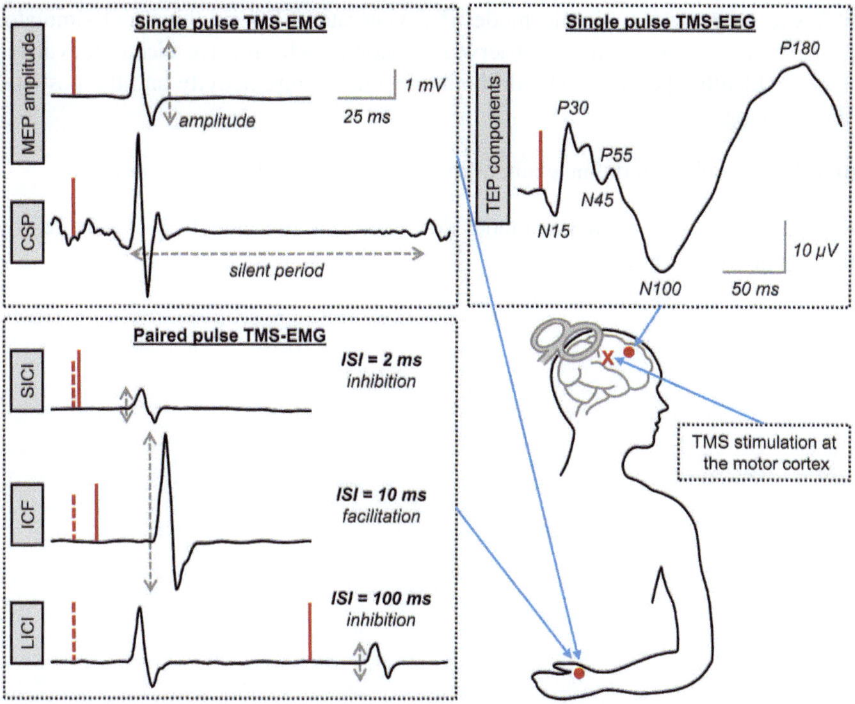

Fig. 10.3 Outcome measures for TMS-EMG and TMS-EEG. Upper panels correspond to the single pulse TMS paradigm and the lower panel to the paired pulse TMS paradigm. Red vertical solid lines: TMS pulse; red vertical dashed lines: conditioning TMS pulse. MEP: motor evoked potential; CSP: cortical silent period; TEP: TMS evoked potential; SICI: short intracortical inhibition; ICF: intracortical facilitation; LICI: long intracortical inhibition and ISI: interstimulus interval. Reprinted from [32], with permission from Elsevier Ireland Inc.

limitations in the measurement technique resulting from stimulation of peripheral sensory and motor nerves in the scalp or the sound generated by the coil once the current passes [24, 98].

Problems

10.1 Can you illustrate how a perturbation by a short electrical stimulus can change the dynamics of a system from a stable equilibrium towards stable limit cycle behaviour? Hint: you can use various of the reduced models introduced earlier in Chap. 4.

10.2 Can you make a simple model that generates EEG rhythms and simulates the generation of a seizure as observed in patients who receive electroconvulsive therapy? Ideally, the model should both generate seizure activity and the postictal EEG slowing.

10.3 How would you define excitability of a neuron or a neuronal assembly?

10.4 Can you make a simple model that simulates the induction of seizures by electroconvulsive therapy?

Correction to: Dynamics of Neural Networks

Correction to:
M. J. A. M. van Putten, *Dynamics of Neural Networks*, https://doi.org/10.1007/978-3-662-61184-5

In the original version of the book, the following corrections have been incorporated:

1. In Chap. 4, Fig. 4.3 has been updated.
2. Links to electronic supplementary material in Chaps. 1, 2, 4, 5, 6, 7, 8, 9 have been included.

The book and the chapters have been updated with the changes.

The updated version of these chapters can be found at
https://doi.org/10.1007/978-3-662-61184-5_1,
https://doi.org/10.1007/978-3-662-61184-5_2,
https://doi.org/10.1007/978-3-662-61184-5_4,
https://doi.org/10.1007/978-3-662-61184-5_5,
https://doi.org/10.1007/978-3-662-61184-5_6,
https://doi.org/10.1007/978-3-662-61184-5_7,
https://doi.org/10.1007/978-3-662-61184-5_8,
https://doi.org/10.1007/978-3-662-61184-5_9 and
https://doi.org/10.1007/978-3-662-61184-5

M. J. A. M. van Putten, *Dynamics of Neural Networks*,
https://doi.org/10.1007/978-3-662-61184-5_11

Appendix A
Software and Programs

We used several software programs in this book. Useful tools for simulations include

- MATLAB https://www.mathworks.com Matlab is a generic software toolbox for many applications, ranging from simulation, signal processing and statistics to finance. Dedicated toolboxes: pplane for phase-plane analysis.
- dfield and pplane Dfield (direction field) and pplane (phase plane) are software programs for the interactive analysis of ordinary differential equations (ODE) and are copyrighted in the name of John C. Polking, Department of Mathematics, Rice University. Both Matlab and java version are available.
- Python https://www.python.org Open source programming language for many applications, including neuronal modeling.
- Neuron https://www.neuron.yale.edu/neuron/ The NEURON simulation environment is used in laboratories and classrooms around the world for building and using computational models of neurons and networks of neurons.
- GENESIS https://www.genesis-sim.org/ GENESIS (the GEneral NEural SImulation System) is a general purpose simulation platform that was developed to support the simulation of neural systems ranging from sub-cellular components and biochemical reactions to complex models of single neurons, simulations of large networks, and system-level models.
- Brian https://www.briansimulator.org/ Brian is a free, open source simulator for spiking neural networks. It is written in the Python programming language and is available on almost all platforms. We believe that a simulator should not only save the time of processors, but also the time of scientists. Brian is therefore designed to be easy to learn and use, highly flexible and easily extensible.
- XPP/XPPAUT https://www.math.pitt.edu/~bard/xpp/xpp.html XPPAUT is a general numerical tool for simulating, animating, and analyzing dynamical systems.
- Simbrain https://www.simbrain.net SIMBRAIN is a free tool for building, running, and analyzing neural-networks (computer simulations of brain circuitry). Simbrain aims to be as visual and easy-to-use as possible.
- Neuronify https://ovilab.net/neuronify/ Neuronify is an educational simulator for neural circuits based on integrate-and-fire neurons.

M. J. A. M. van Putten, *Dynamics of Neural Networks*,
https://doi.org/10.1007/978-3-662-61184-5

- NEST https://www.nest-simulator.org/ NEST is a simulator for spiking neural network models that focuses on the dynamics, size and structure of neural systems rather than on the exact morphology of individual neurons. The development of NEST is coordinated by the NEST Initiative. NEST is ideal for networks of spiking neurons of any size, for example: (i) Models of information processing e.g. in the visual or auditory cortex of mammals; (ii) Models of network activity dynamics, e.g. laminar cortical networks or balanced random networks or (iii) Models of learning and plasticity.

A listing of various programs used in each chapter is available on the website.

Appendix B
Solutions to the Exercises

Chapter 1

1.1 Typical values for the Nernst potentials for sodium, potassium and calcium at T = 37°C are (Table B.1). Note that $z = +2$ for Ca^{2+}.

1.2 Use the rule that tells you how to define an average conductance.

1.3 This is of the order of one tot two minutes. For additional reading, see [137, 145]. These papers present and discuss an interesting experimental observation in animals, and show simulations of a particular phenomenon, now known as "The Wave of Death".

1.4 The activation function is shown in Fig. B.1.

1.5 See Izhikevich, Chap. 2 [59].

1.6 Use $n_\infty = \alpha_n/(\alpha_n + \beta_n)$ and $\tau_n = 1/(\alpha_n + \beta_n)$, and insert these into (1.24).

1.7

a. Voltage-gated ion channels: soma, dendrites and axons of all neurons. Responsible for the generation of action potentials. Examples include voltage-gated Na^+ or K^+ channels.
b. Ligand-gated ion channels, present on postsynaptic membranes. Activated by neurotransmitters, such as acetylcholine or dopamine.
c. Stretch/pressure/temperature receptors. Present on dendrites of neurons of the peripheral nervous system, turning physical stimuli into an electrical signal (transduction).

1.8 You should use the equation describing the behavior of neurons as an RC circuit (where the C is in parallel with the resistor R and voltage source with value $V(0)$)

M. J. A. M. van Putten, *Dynamics of Neural Networks*,
https://doi.org/10.1007/978-3-662-61184-5

Table B.1 Ion concentrations across the cell membrane

Ion	Intracellular (mM)	Extracellular (mM)	Nernst potential (mV)
sodium (Na^+)	10	140	$E_{Na} = +71.1$
potassium (K^+)	140	4	$E_K = -95.7$
chloride (Cl^-)	5	110	$E_{Cl} = -83.2$
calcium (Ca^{2+})	0.1	4	$E_{Ca} = +143$

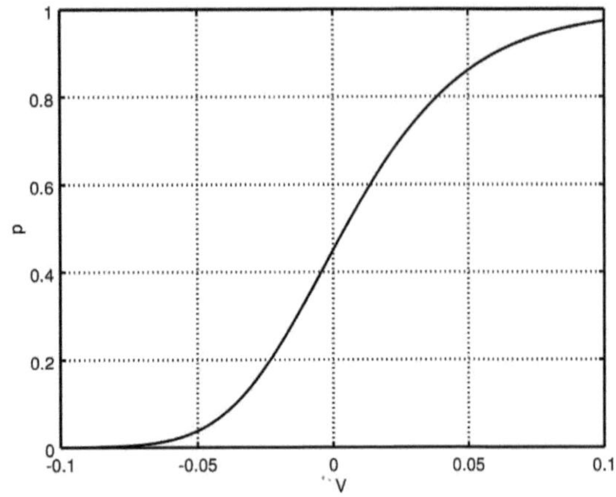

Fig. B.1 Activation function using $b = 0$, $m = 5/(5 + \exp(0.1 - 35 \cdot V))$ and $a = 4$ in (1.21)

(see Fig. 1.3, omitting the potassium and sodium channels with their batteries and setting $E_L = V(0)$.) and specifically the formula for charging a membrane, given by

$$V(t) = (V_0 - V_\infty)e^{-t/\tau} + V_\infty$$

where $\tau = r_m \cdot c_m$. Using the values given, τ=1 µF/cm$^2 \cdot$ 2000 ohms cm^2 = 2 ms. This results in V = 5.04 mV at t = 2 ms.

Use the equation that describes the discharge of a membrane following removal of a stimulus

$$V(t) = V_0 e^{-t/\tau}.$$

This results in V = 8 mV $e^{-4/2} \approx$ 8 mV $\cdot$ 0.14 = 1.08 mV above the rest level.

1.9 At the peak of the action potential, $I = 0$, and $I_{Na} = -I_K$. Substituting equations for I_{Na} and I_K, $V_m = (g_{Na}E_{Na} + g_K E_K)/(g_{Na} + g_k) = 40.3$ mV.

1.10 Use

$$I_{ion} = g_{ion}(V_m - E_{ion})$$

Since the current $I = 0$ and conductance is not, V_m must equal E_{ion}. Using the Nernst equation, you find that $E_{ion} = -84$ mV, so $V_m = -84$ mV.

1.11 Lower the extracellular sodium concentration to such a value that its Nernst potential equals the command voltage (i.e. the voltage set in the voltage-clamp experiment). In those circumstances, the currents are mainly the result of the potassium channels, since at the command voltage the sodium current is zero, according to $I_{\mathrm{Na}} = g_{\mathrm{Na}}(V_{\mathrm{command}} - V_{\mathrm{Na}}) = g_{\mathrm{Na}} \cdot 0$.

1.12 First, use the voltage clamp to obtain $\mathrm{I}_{\mathrm{total}}$ at a single V_m. This is essentially $\mathrm{I}_{\mathrm{Na}} + \mathrm{I}_K + \mathrm{I}_{\mathrm{Ca}}$. Then, remove one current at a time from the total current and record. I_{Na} can be removed with TTX leaving $\mathrm{I}_{\mathrm{total}} = I_K + I_{\mathrm{Ca}}$. I_K can be removed by changing the external concentration of K^+ with potassium salt, shifting E_k to the command voltage. With no driving force on K^+ ions the $\mathrm{I}_{\mathrm{total}} = I_{\mathrm{Na}} + I_{\mathrm{Ca}}$. Subtracting both of these total currents from the original $\mathrm{I}_{\mathrm{total}}$ leaves us with I_{Ca} from which we can calculate the Ca conductance, g_{Ca}. Finally, repeat these recordings at different V_m and calculate the full range of g_{Ca}.

1.13

a. The patch clamp technique. This allowed the measurement of tiny quantal currents, as if a channel was flickering open and closed. The currents are of the order of pA: single channels have small currents.
b. sign of the current, transient (Na) or persistent (K) current, threshold of significant channel opening.

1.14 –

1.15 In situations where the Nernst potential of chloride, controlled by the GABA-b receptor, is above (less negative) than the present membrane potential, opening of the GABA-b-receptor will tend to change the membrane potential towards the Nernst potential of chloride. Note, that this is still an inhibitory current. Compare this line of reasoning with (1.10).

1.16

a. For a realistic value of a membrane thickness of 5-6 nm, and a voltage of -70 mV, we find

$$E = \frac{-dV}{dx} = -\frac{-70 \cdot 10^{-3} V}{6 \cdot 10^{-9}\mathrm{m}} = 1.17 \cdot 10^7 \; V/\mathrm{m}$$

b. Assume the membrane is a parallel plate capacitor, with area A and specific capacitance $c_m = 1$ μF/cm^2. The surface charge, therefore, is, using $Q = C \cdot V = 10^{-6} \cdot 70 \cdot 10^{-3} = 7 \cdot 10^{-4}$ C/m^2.
c. For the force between a parallel plate capacitor (assuming infinite dimensions[1]), it holds that

$$F = k\varepsilon_o \frac{AV^2}{2d^2}.$$

[1]For finite size, a correction is needed. For circular plates, one should multiply (B.1) with $(1 + 2d/D)$ where D is the diameter of the plate and d the distance between the plates.

The pressure P exerted on the "dielectricum" is then

$$P = k\varepsilon_o \frac{V^2}{2d^2}.$$

Setting the dielectric constant k=7, and using $\varepsilon_o = 8.85 \cdot 10^{-12}$ F/m we find for the pressure P,

$$P = 7 \cdot 8.85 \cdot 10^{-12} \cdot \frac{(70 \cdot 10^{-3})^2}{2 \cdot (6 \cdot 10^{-9})^2} \approx 4216 \text{ N/m}^2.$$

d. Heat is generated while ionic currents pass the cell membrane, as the resistance is finite. The cause for the change in diameter (local volume change) is still being debated. Can you explain the change in diameter from the pressure and the properties of the cell membrane alone, or are other mechanisms needed?

1.17

a. You should observe that if you increase the extracellular concentration more and more, the neuron can start spiking even without an external current I!
b. Clinical conditions where the extracellular potassium is increased (hyperkalemia) include acute kidney failure, chronic renal failure and dehydration.

1.18 –

Chapter 2

2.1 Yes, this is possible. For instance, if the Nernst potential of chloride is above the resting membrane potential, opening of the GABA-receptor will result in an increase in the membrane potential, but the effect is still inhibitory.

2.2 Take the derivative of (2.10) or (2.11) and set this equal to zero to obtain

$$t_{\max} = \frac{\ln(\tau_1/\tau_2) \cdot \tau_1 \cdot \tau_2}{\tau_1 - \tau_2}.$$

Note, that $\tau_1 > \tau_2$, as stated in the text.

2.3 If you take the derivative, it is straightforward to prove that the maximum value is reached at $t = \tau$.

2.4

a. The circuit will need to be expanded with a *synaptic battery* with associated reversal potential E_{syn} with a series conductance, g_{syn}.

b. Various other elements are first to be removed, only keeping the synaptic battery and conductance and the leak conductance.
c. For a rectangular pulse with an amplitude of g_{syn} and a duration of t_{syn}, we now find for the membrane potential as a function of time

$$V_m(t) = \frac{g_{\text{syn}}}{g_{\text{syn}} + g_{\text{leak}}} E_{\text{syn}}(1 - e^{-t/\tau})$$

for $0 \leq t \leq t_{\text{syn}}$ and $\tau = C_m/(g_{\text{syn}} + g_{\text{leak}})$.
d. As during synaptic activity, g_{syn}, is larger than in rest, the time constant during opening is different than the time constant during closing. In this latter situation, g_{syn} is relatively small and the time constant $\tau \approx C_m/g_{\text{leak}}$. Therefore, during application of an external current, the membrane potential will rise *slower* than when synaptic channels are opened. The larger g_{syn} the faster $V_m(t)$ develops towards its maximal value.

2.5 The predominant excitatory neurotransmitter is the amino acid glutamate, whereas in the peripheral nervous system it is acetylcholine. Glutamate-sensitive receptors in the post-synaptic membrane can be subdivided into two major types, namely NMDA and AMPA. At an AMPA receptor, the postsynaptic channels open very rapidly, with an exponential decay of about 1 ms. The NMDA receptor operates about 10 times slower. The most common inhibitory neurotransmitter in the central nervous system appears to be GABA. There are two major forms of postsynaptic GABA receptors termed A and B. The GABA_A receptor opens channels selective to chloride ions, and the conductance change is fast, rising within 1 ms, and decaying withing 10-20 ms. The GABA_B receptors are at least 10 times slower, and open channels selective for potassium ions.

2.6 If the synaptic input stops, the membrane decay time constant can be approximated by

$$\tau_m = C_m/g_{\text{rest}}.$$

assuming g_{syn} is now very small (channels are closed). If synaptic input is e.g. approximated by an alpha function with time constant $\tau_\alpha < \tau_m$ the response of the cell membrane voltage after the synaptic input is essentially defined by τ_m.

2.7 You should find values for the time constant of approximately 1–2 ms (baseline) and 5–7 ms (day 36).

2.8

a. There are various types of Myasthenia Gravis. In a significant proportion of patients, antibodies are formed against the acetylcholine receptor, that interfere with its function. This reduces $\bar{g}$ in the synaptic transfer functions.
b. Lowering the temperature may limit the action of enzymes. Acetylcholine is partially removed from the synaptic cleft by enzymatic breakdown by acetylcholinesterase. Lowering the temperature, therefore, will increase the concentration of acetylcholine in the synaptic cleft, resulting in an increase in $\bar{g}$.

c. Essentially, $\bar{g}$ is increased. Remember, that in principle two mechanisms can be employed to increase the maximum synaptic conductance. First by increasing the amount of neurotransmitter (if this is the limiting value) and secondly by increasing the number of ligand-gated channels. This is clinically applied, too, e.g. by treating patients with prednison or plasmapheresis.

Chapter 3

3.1

a. $x(t) = \frac{1}{1-t}$. Note that $x \to \infty$ at $t = 1$
b. $x(t) = \tan t$.
c. The solution is given by $x(t) = e^{-t/\tau}$. The time where $x(t) = 1/2$ is given by $T_{1/2} = \tau \ln 2$.

3.2

a. $x\,(t) = \mathrm{e}^{-\sin(t)}$;
b. $x\,(t) = (\cos\,(t))^{-1}$.

3.3 –

3.4

b. $x\,(t) = C\mathrm{e}^{-2t} - \frac{\omega\cos(\omega t) - 2\sin(\omega t)}{\omega^2+4}$.

3.5 $x\,(t) = -\mathrm{e}^{-t}$. If $t \to \infty$ then $x = 0$.

3.6 Fixed points: $x* = k\pi$. To evaluate if these are stable or unstable, we take the derivative of f(x) at $x*$, which is $\cos(k\pi)$. Therefore, $x*$ is stable if k is odd and unstable otherwise.

3.7

b. $x(t) = \frac{x_o}{x_o(1-e^{-t})+e^{-t}}$
c. The escape time $T = \ln\left((x_0 - 1)/x_0\right)$ assuming that $(x_0 - 1)/x_0 > 0$.

3.8 At the stable equilibrium $N = K$.

3.9

a. Set $f(x) = x(1-x)(x-a)$. Equilibria are $x = 0,\ x = 1,\ x = a$. Take the derivative of $f(x)$ and evaluating at the fixed points shows that for the values of $a \in (0, 1)$ $x = 0$ and $x = 1$ are stable fixed points and $x = a$ is an unstable fixed point.

b. Set $f(x) = x - x^3$. Equilibria are $x = 0,\ x = -1$ and $x = 1$. Using $f'(x) = 1 - 3x^2$ and evaluating at these values shows that $f'(0) = 1$, and $f'(-1) = f'(1) = -2$. Therefore, $x = 0$ is an unstable equilibrium and the other two equilibria are stable.

3.10 For $a > 0$, the origin is a stable fixed point (attractor). For $a < 0$, the origin is unstable (a repellor). If a=0, a bifurcation occurs at the origin $x = 0$. This bifurcation is known as an *exchange of stability bifurcation*.

3.11

a. We solve $-x + x^2 + \lambda = 0$ and $-1 + 2x = 0$. It follows that $\lambda = 1/4$ and $x = 1/2$.
b. We solve $-x + x^3 + \lambda = 0$ and $-1 + 3x^2 = 0$. It follows that $x = \pm 1/3\sqrt{3}$ and $\lambda = \pm 2/9\sqrt{3}$.
c. $x = \pm 1$ and $\lambda = \pm 2$.
d. $x = 0$ and $\lambda = 0$.
e. $x = -1$ and $\lambda = \pm 1$.

3.12 –

3.13 When $a > 0$ equilibria are $x = 0$ and $x = \pm\sqrt{a}$. When $a \leq 0$, $x = 0$ is the only equilibrium point.

Chapter 4

4.1

(a) $-3 \begin{pmatrix} -2 \\ 1 \end{pmatrix}, 2 \begin{pmatrix} 1 \\ -3 \end{pmatrix}$

(b) $-4 \pm 2i \begin{pmatrix} 1 \\ \mp i \end{pmatrix}$

(c) $3 \begin{pmatrix} 1 \\ -\frac{7}{2} \end{pmatrix}, -4 \begin{pmatrix} 1 \\ 0 \end{pmatrix}$

4.2

(a). $\lambda_1 = -2$, $\lambda_2 = -1$. As the $\det(A) = 2$, the $tr(A) = -3$ and $4 \cdot \det(A) - tr(A)^2 < 0$, the origin is a stable node (use Fig. 4.2).
(b). $\lambda_1 = -2$, $\lambda_2 = 1$. As the $\det(A) < 0$ the origin is a saddle point.
(c). $\lambda_1 = -1 + i$, $\lambda_2 = -1 - i$. The origin is a stable spiral. The direction fields are shown in Fig. B.2.

4.3 $\frac{d}{dt} e^{\lambda t} v = \lambda e^{\lambda t} v = e^{\lambda t} A v.$

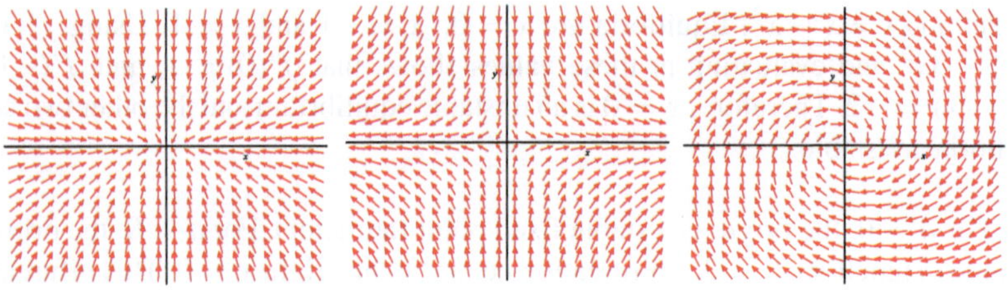

Fig. B.2 In all three cases all orbits converge to the origin. The coordinate axes are invariant under the flow in case **a** (left) and **b** (middle). In case **c** (right), the orbits spiral towards the origin

4.4 There is only one eigenvector for the repeated eigenvalue $\lambda = 2$: $\begin{pmatrix} 1 \\ -2 \end{pmatrix}$. This also implies that we have not found a general solution for $\dot{\boldsymbol{x}} = \boldsymbol{A}\boldsymbol{x}$. Recall, that we need two fundamental solutions to write down the general solution for a second order linear differential equation. It can be shown that in this particular case the general solution is given by $x(t) = c_1 \begin{pmatrix} 1 \\ -1 \end{pmatrix} e^{2t} + c_2 \left[\begin{pmatrix} 1 \\ -1 \end{pmatrix} te^{2t} + \begin{pmatrix} 0 \\ -1 \end{pmatrix} e^{2t} \right]$. Check that as $t \to \infty$, $\boldsymbol{x}$ is asymptotic to the line $x_2 = -x_1$, determined by the first eigenvector. The origin is an *improper* or *degenerate* node and is unstable.

4.5 No.

4.6 $c_1 e^{-3t} \begin{pmatrix} -2 \\ 1 \end{pmatrix} + c_2 e^{2t} \begin{pmatrix} 1 \\ -3 \end{pmatrix}$. $c_1 = -\frac{4}{5}, c_2 = -\frac{3}{5}$.

4.7

(a) $\lambda_1 = 0$; $\lambda_2 = 3$.
(b) $\xi_1 = (-0.8944,\ 0.4472)$; $\xi_2 = (-0.7071,\ -0.7071)$
(c) See Fig. B.3

4.8

(a) $b > 2$.
(b) $\det(A) < 0$, where A is the linearization at the fixed point.
(c) $b > \frac{5}{2}$.

4.9 Trajectories of autonomous systems cannot cross: if the future of a similar position in the phase plane, that is visited on different moments in time, could be a choice, time dependence exists. But autonomous systems do not have time dependence, by construction. Therefore, trajectories can never cross (except for the fixed points, where the system is at rest).

4.10 The origin (0.0) is a saddle. The other fixed point at $(1, -1)$ is a center. Check this with the determinant and trace of the linearized system at this equilibrium.

4.11 The trajectories are shown in Fig. B.4.

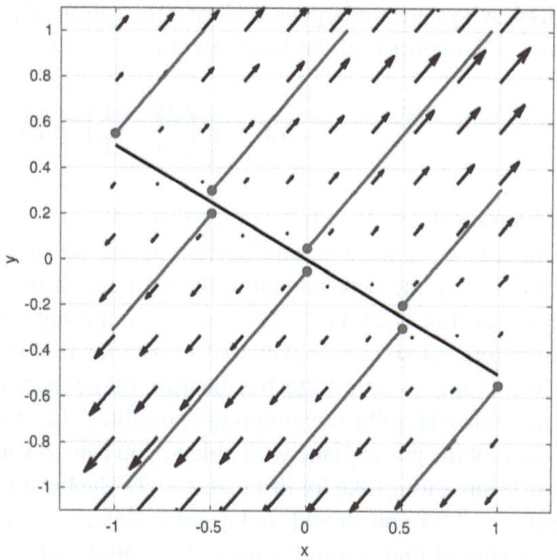

Fig. B.3 Lines of unstable fixed points, corresponding to a system with zero determinant and positive trace. Here, the trace $\tau = 3$

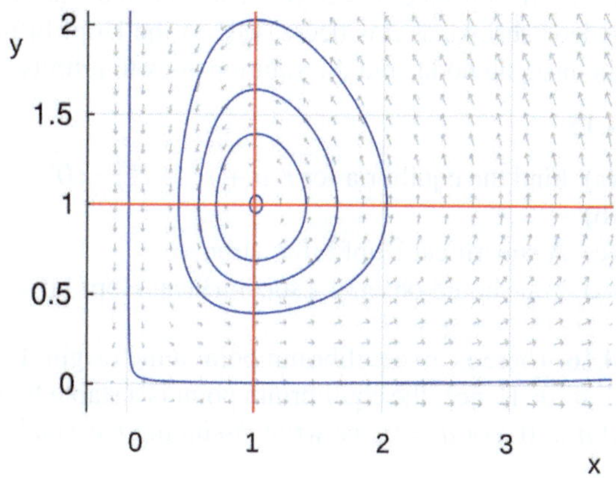

Fig. B.4 Direction field and orbits around the fixed point (1, 1). Both populations oscillate in size, but neither becomes extinct. Can you argue why the direction of the orbits are counter clockwise?

4.12 $x = a,\ y = a - \frac{1}{3}a^3$. (b) $A = \begin{pmatrix} 1 - a^2 & -1 \\ b & 0 \end{pmatrix}$. Equilibrium stable if $|a| > 1$ and unstable if $|a| < 1$.

4.13 The eigenvalues (λ_1, λ_2) for $a = -2$ are $(2, -1)$ and $(-2, -1)$ (two fixed points exist). For $a = 0$ the eigenvalue is $(-1, 0)$. If $a = 2$ the eigenvalues are given by $(2, -1)$ and $(-2, -1)$ for each fixed point. For $a = -2$ the initial fixed points were given by $(-2, 0)$ and $(0, 0)$, while for $a = 2$ the fixed points are $(0, 0)$ and $(2, 0)$. At the bifurcation, the fixed point is $(0, 0)$). Recall that in a transcritical bifurcation the fixed points are not destroyed, but their stability is exchanged.

4.14 A system undergoes a Hopf bifurcation if the equilibria cross the imaginary axis, i.e. the trace of the Jacobian at the equilibrium should equal zero, i.e.

$$\mathrm{tr}\boldsymbol{J} = \mathrm{tr}\begin{pmatrix} 2V & -1 \\ b & -1 \end{pmatrix} = 2V - 1 = 0$$

which results in $V = 0.5$. As for this value, we must also satisfy $\dot{V} = 0$ and $\dot{u} = 0$ (otherwise we have no equilibrium at all) we find $b = \frac{1}{2} + 2I$. Further, for a Hopf bifurcation, the determinant at the bifurcation point $V = 0.5$ should be larger than zero (cf. Fig. 4.2), i.e. $-2V + b > 0$, implying that $b > 1$. The input current must therefore satisfy $I > \frac{1}{4}$. Check for instance that if you set $I = 0$ the equilibrium at the origin is a stable node, but the trace of the Jacobian at the origin $\tau = -1 \neq 0$, and therefore a Hopf bifurcation is not possible. The other equilibrium is a saddle node. If you plot the phase plane with pplane you can now also evaluate the characteristics of the bifurcation. Take for instance $I = 1$. The bifurcation occurs at $(0.5, 0.5 \cdot b$. If you set $b = 2.55$ you should find a spiral sink at $(V, u) \approx (0.48, 1.23)$ and at $b = 2.45$ you should find a spiral source. The spiral sink is surrounded by an unstable limit cycle. If you take $b = 3$ you will find that the unstable limit cycle is larger, and indeed shrinks as b is decreased. At the Hopf bifurcation the stable node becomes an unstable node, and the solution goes to infinity. See Fig. B.5.

4.15

(a) Find the equilibria for $r(a + r^2 - r^4) = 0$.
(b) –
(c) A subcritical Hopf bifurcation
(d) You should observe a supercritical Hopf bifurcation now.

4.16 There is an equilibrium point a the origin. For small values of r, it holds that $\dot{r} \approx ar$. Hence, the equilibrium point is a stable focus if $a < 0$ and an unstable focus if $a > 0$. For $a = 0$, we write, assuming r is small

$$\dot{r} = 2r^3 - r^5 \approx 2r^3 > 0.$$

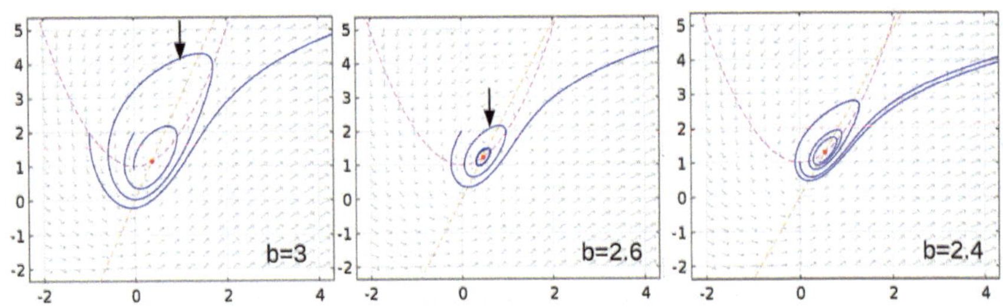

Fig. B.5 Phase plane with $I = 1$ and b as indicated. Note the shrinkage of the unstable limit cycle around the stable node, with the subcritical Hopf into an unstable node, where all solutions now disappear to infinity

This implies that the origin is unstable. Now that we have defined characteristics of the stability of the origin, we try to explicitly find limit cycles that are characterized by $\dot{r} = 0$. We can factorise $\dot{r} = ar + 2r^2 = r^5$ as

$$\dot{r} = r\left(r^2 - 1 - \sqrt{1+a}\right)\left(r^2 - 1 + \sqrt{1+a}\right)$$

if $-a < a < 0$ then $1 \pm \sqrt{1+a}$ is real and positive. We can now further factorise and obtain

$$\dot{r} = r\left(r - \sqrt{1+\sqrt{1+a}}\right)\left(r + \sqrt{1+\sqrt{1+a}}\right)\left(r - \sqrt{1-\sqrt{1+a}}\right)\left(r + \sqrt{1-\sqrt{1a}}\right).$$

As the radius $r > 0$ we obtain to limit cycles with radius $r_1 = \sqrt{1-\sqrt{1+a}}$ and $r_2 = \sqrt{(1+\sqrt{1+a}}$. If you examine the signs of the terms in the factorisation, if follows that r_1 is unstable and r_2 is stable.

If the parameter $a > 0$ then $1 + \sqrt{1+a}$ is real and positive but $1 = \sqrt{1+a}$ is real and negative, so factorisation of (B.4) then becomes

$$\dot{r} = r\left(r - \sqrt{1+\sqrt{1+a}}\right)\left(r + \sqrt{1+\sqrt{1+a}}\right)\left(r^2 + \sqrt{1+a} - 1\right)$$

resulting in a single stable (check the signs of the terms) limit cycle of radius $r_2 = \sqrt{1+\sqrt{1+a}}$.

Summarizing: if $= 1 < a < 0$, the origin is a stable focus surrounded by an unstable limit cycle of radius $r_1 = \sqrt{1-\sqrt{1+a}}$ and a stable limit cycle of radius $r_2 = \sqrt{(1+\sqrt{1+a}}$. If $a \geq 0$ the origin is an unstable focus surrounded by a stable limit cycle of radius $r_2 = \sqrt{1+\sqrt{1+a}}$. Therefore, the bifurcation is a subcritical Hopf. In this example, there exists a stable limit cycle of radius $r = 1$ as the bifurcation occurs, and the system will 'jump' to this stable limit cycle.

4.17 In a supercritical Hopf bifurcation, a stable spiral becomes unstable, but it is surrounded by a small stable limit cycle, keeping the system relatively close to its initial stable state. In a subcritical Hopf bifurcation, however, the stable spiral becomes unstable without emergence of a stable limit cycle, and the trajectories will jump to a distant attractor, that could be far from the initial fixed point. If such a bifurcation would occur in for instance a mechanical construction, severe structural damage may occur. In a biological system, super- and subcritical Hopf bifurcations occur in models for the generation of spike-wave discharges in epilepsy in, see e.g. [44].

4.18 –

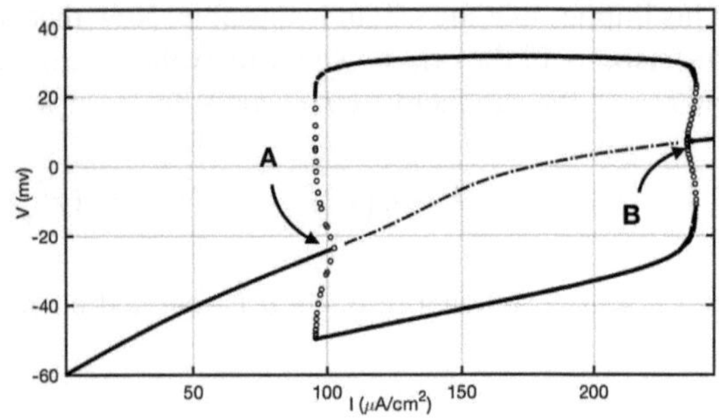

Fig. B.6 Bifurcation diagram of the Morris-Lecar model with parameter values S1. Both at A and B, a subcritical Hopf bifurcation occurs

4.19

a. Using parameters S1, the bifurcation diagram is sketched in Fig. B.6. There are two *subcritical Hopf bifurcations*. The first for a current of about 101.8μA/cm^2, the second one is for a higher current of about 220 μA/cm^2. Type II behavior occurs: a jump at the bifurcation to a finite firing frequency.
b. If parameters S2 are used, the bifurcation diagram is very different. The bifurcation is a Saddle-Node-off-Limit-Cycle bifurcation, and Type I behavior is present, as the frequency of the oscillations can be arbitrarily small.

4.20

a. From $\frac{dh}{dV} = 0$ it follows that

$$v = \frac{1}{3}(a+1) \pm \frac{1}{3\gamma}\sqrt{(a^2 - a + 1)\gamma^2 - 3\gamma}$$

   To have a solution (which then corresponds to a local extremum) we therefore need that $(a^2 - a + 1)\gamma - 3 \geq 0$ (as $\gamma \geq 0$).
b. The fixed point exists and is unique as h is strictly increasing with range $\mathscr{R}$.
c. The linearization at the fixed point $(\bar{V}, \bar{w})$ is given by

$$\begin{pmatrix} -3\bar{V}^2 + 2(a+1)\bar{V} - a & -1 \\ \varepsilon & -\gamma\varepsilon \end{pmatrix}.$$

   Hence the condition is $-3\bar{V}^2 + 2(a+1)\bar{V} - a - \gamma\varepsilon = 0$.

4.21 Hysteresis.

Chapter 5

5.1 –

5.2 –

5.3 –

5.4

$$f \approx \frac{I + E_{\text{rest}} - V_{\text{threshold}}}{\tau_m (V_{\text{threshold}} - V_{\text{reset}})}$$

5.5 –

Chapter 6

6.1 GABA activates the GABA receptor, which will result in chloride ions entering the neuron (current source), whereas the glutamate receptor will mainly transport sodium ions inside the neuron, acting as a current sink. Remember that positive charges are considered as positive currents, and a current source transports positive charges *from the neuron into the extracellular space*.

6.2 $\mathbf{p} = 0$ as $a = 0$ in neurons with a closed field.

6.3

a 1.41 μV. Note that as the measurement position is close to the current source, you cannot use the approximation (6.6), but you should use (6.4). The difference is about 0.4 μV.
b 0.01 μV. This voltage is too small to allow reliable measurement at the scalp, as the noise level of current instrumentation is of the order of 0.1 μV.

6.4 Hint: start with (6.9) and recall that the Taylor series of $\sqrt{1+x}$ centered at $x = 0$ can be expressed as

$$1 + \frac{x}{2} - \frac{x^2}{8} + \frac{x^3}{16} - \cdots$$

6.5 0.85 μV.

6.6 As the common-mode rejection ratio (CMRR) of the EEG amplifier is finite, the patient and amplifier should be connected to the same ground.

6.7 The sensitivity is approximately 30–40%, depending on the epilepsy type. One may increase the sensitivity by longer recordings or recordings during sleep.

6.8 Absence seizures (and some other generalized epilepsies) are exacerbated by phenytoin and carbamazepine. See e.g. [119] for an explanation.

Chapter 7

7.1 The unit of ν is mVs. Furthermore, the unit of h and Q_{in} is 1/s and since the convolution $*$ is an integral over time (in units of s), the unit of $h * Q_{in}$ is 1/s. Therefore, the unit of $\nu h * Q_{in}$ is mV.

7.2 Write $\beta = \alpha + \varepsilon$ for certain $\varepsilon > 0$. Then

$$
\begin{aligned}
\lim_{\beta \to \alpha} h(t) &= \lim_{\varepsilon \to 0} \frac{\alpha(\alpha+\varepsilon)}{\varepsilon}(1-e^{-\varepsilon t})e^{-\alpha t} \\
&= \alpha^2 e^{-\alpha t} \lim_{\varepsilon \to 0} \frac{1-e^{-\varepsilon t}}{\varepsilon} && \text{because 2nd term vanishes as } \varepsilon \to 0 \\
&= \alpha^2 e^{-\alpha t} \lim_{\varepsilon \to 0} t e^{-\varepsilon t} && \text{l'Hôpital applied} \\
&= \alpha^2 t e^{-\alpha t}.
\end{aligned}
$$

7.3 See Chap. 2.

7.4 A decrease of receptor de-activation can be modeled by a decrease of α. However, since a change in α influences the maximal height to h, we have to compensate for this (since propofol leaves the maximal height unaltered). Thus, we define a new synaptic response $g(t)$ which is defined as

$$
g(t) = \frac{H}{N(\alpha, \beta)} h(t)
$$

where $N(\alpha, \beta)$ is the maximal height of $h(t)$ (as a function of α and β). It follows that H is the maximal height of $g(t)$. The normalization constant $N(\alpha, \beta)$ is found by setting the time-derivate of $h(t)$ to zero, which shows that h is maximal at $t = t^*$, where

$$
t^* = \frac{\ln(\beta/\alpha)}{\beta - \alpha}
$$

which gives

$$
N(\alpha, \beta) = h(t^*) = \frac{\alpha\beta}{\beta - \alpha}\left(e^{-\alpha \frac{\ln(\beta/\alpha)}{\beta-\alpha}} - e^{-\beta \frac{\ln(\beta/\alpha)}{\beta-\alpha}}\right)
$$

7.5

$$
\begin{aligned}
S'(V) &= \frac{Q_{max}}{\sigma} \frac{e^{-(V-\theta)/\sigma}}{(1+e^{-(V-\theta)/\sigma})^2} \\
&= \frac{S(V)}{\sigma}\left(\frac{1}{1+e^{-(V-\theta)/\sigma}} e^{-(V-\theta)/\sigma}\right) \\
&= \frac{S(V)}{\sigma}\left(1 - \frac{S(V)}{Q_{max}}\right).
\end{aligned}
$$

In the last identity we have used that $e^{-(V-\theta)/\sigma} = \frac{Q_{max}}{S(V)} - 1$.

7.6 in the Laplace domain, the convolution in (7.3) is given by the product of the Laplace transforms of each function, i.e. $\mathscr{L}\{V(t)\} = V(s) = \mathscr{L}\{\nu h(t)\}\mathscr{L}\{Q_{in}(t)\} = \nu\mathscr{L}\{h(t)\}\mathscr{L}\{Q_{in}(t)\} = \nu H(s)Q_{in}(s)$. Use $H(s) = \frac{\alpha\beta}{(s+\alpha)(s+\beta)}$, which follows from the elementary properties of the Laplace transformation. Now substitute $H(s)$ in the Laplace transform of (7.46)

$$s^2V(s) + (\alpha+\beta)sV(s) + \alpha\beta V(s) = \alpha\beta\nu Q_{in}(s)$$

and you will find that the equality holds.

7.7 Substituting $V(t)$ and $Q_{in}(t)$ into (7.46) gives

$$(-\omega_0^2 + (\alpha+\beta)i\omega_0 + \alpha\beta)Ae^{i\omega_0 t} = \delta e^{i\omega_0 t},$$

from which we find that

$$A = \frac{\alpha\beta\nu\delta}{(i\omega_0+\alpha)(i\omega_0+\beta)}.$$

It follows that $|A|^2 = P_{EEG}(\omega_0)$.

7.8 The effect of a peptide neurotoxin is that it decreases ν until it equals zero. It follows that the membrane potential of the neural mass becomes flat (despite incoming fluctuations) and equal to zero (which equals the mean resting potential of the neural mass).

7.9 We cannot estimate δ since P_{EEG} is proportional to both δ and ν. We also cannot estimate α and β since P_{EEG} is symmetrical in α and β (we can exchange α and β without changing the power spectrum). If $\beta = 4\alpha$ then the denominator of P_{EEG} can be re-written as

$$\omega^4 + 17\alpha^2\omega^2 + 16\alpha^4.$$

Since the term $17\alpha^2\omega^2$ determines the shape of this polynomial, we can estimate this term and hence α.

7.10 This is similar to Exercise 7.6.

7.11 The steady states correspond to the intersections between the line $l(V) = -\nu q/\mu + V/\mu$ and $S(V)$. If $\mu < 0$, the line l is decreasing. Since S is increasing, they intersect in one point.

7.12

a The mass has exactly one steady state if the slope of l is larger than the maximal slope of S (see Fig. 7.9). Since the slope of S is maximal in the point $V = \theta$, it follows that this is equivalent to $dl/dV = 1/\mu$.

b Using Exercise 7.5 and the fact that $dl/dV > dS/dV(\theta)$ we find that the condition $dl/dV > dS/dV(\theta)$ is equivalent to $\mu < 4\sigma/Q_{max}$.

c With $\dot{x_1} = \dot{x_2} = 0$ we find that $x_2 = \dot{V} = 0$. So, at the equilibrium the voltage of the neural mass is constant. From $\dot{x_2} = 0$ we arrive at $-x_1 + \nu q + \mu \frac{Q_{\max}}{1+e^{-(x_1-\theta)/\sigma}} = 0$.

d plot $f(x) = -x + \nu q + \mu \frac{Q_{\max}}{1+e^{-(x-\theta)/\sigma}}$ and explore the equilibria. You will find that if the condition for three equilibria is satisfied $f(x)$ will have a W-shape. This makes three intersections with the line $y = 0$ possible if q is sufficiently large.

7.13 If q is very large, then it follows from (22) that V^* is also very large (because the term $\mu S(V^*) > 0$). Since the steady-state firing rate Q^* is related to V^* by $Q^* = S(V^*)$ and S is an increasing function, it follows that Q^* will approach Q_{max}, which is the maximum of S. A similar argument shows that if q becomes very negative, Q^* will approach 0.

7.14 It increases.

7.15 This follows directly by solving the second-order polynomial $(\lambda + \alpha)(\lambda + \beta) - \alpha\beta G$ for λ.

7.16 The mass destabilizes exactly when $G = 1$. From (25) we know that $G = \mu S'(V^*)$. Using exercise (5) we find that

$$1 = \mu \frac{Q^*}{\sigma}\left(1 - \frac{Q^*}{Q_{max}}\right),$$

$$(Q^*)^2 - Q_{max} Q^* + \frac{\sigma Q_{\max}}{\mu} = 0,$$

and solved for Q^* to obtain the values of Q^*.

7.17 When $G = 0$, the resonances are given by $\lambda = -\alpha, -\beta$. When G gets more negative, the term $(\alpha - \beta)^2 + 4\alpha\beta G$ approaches zero, hence the resonances approach each other, as can be seen from Exercise 7.15. When this term reaches zero, the resonances collide and equal $-(\alpha + \beta)/2$. This happens when $(\alpha - \beta)^2 + 4\alpha\beta G$, which is equivalent to $G = -(\alpha - \beta)^2/4\alpha\beta$.

7.18 If $G = 0$ the EEG power spectrum reduces to

$$P_{EEG}(\omega) = |\frac{\alpha\beta\nu\sigma}{(i\omega + \alpha)(i\omega + \beta)}| = \frac{(\alpha\beta\nu\sigma)^2}{|i\omega + \alpha|^2 |i\omega + \beta|^2} = \frac{(\alpha\beta\nu\sigma)^2}{(\omega^2 + \alpha^2)(\omega^2 + \beta^2)}$$

7.19 We re-write the denominator of the EEG power spectrum as follows:

$$\begin{aligned} |(i\omega + \alpha)(i\omega + \beta) - G|^2 &= |-(\omega^2 + \alpha\beta - G) + i\omega(\alpha + \beta)|^2 \\ &= (\omega^2 + \alpha\beta - G)^2 + \omega^2(\alpha + \beta)^2 \\ &= (\alpha^2 + \omega^2)(\beta^2 + \omega^2) + 2(\omega^2 - \alpha\beta)G + G^2. \end{aligned}$$

7.20 Similar to Exercise 7.6.

7.21. The steady states correspond to the intersections of the line $l(V_e) = -\nu q/\mu_{ie}$ and the function $V_e \to S(\mu_{ei} S(V_e))$. Since $\mu_{ie} < 0, l$ is decreasing. Moreover, since $V_e \to S(\mu_{ei} S(V_e))$ is a concatenation of two increasing functions, it is increasing. Since a decreasing and an increasing function intersect in one point, there is one steady state.

7.22 Blocking of the excitatory synapses is modeled by setting $\mu_{ei} = 0$, from which it follows that $G_{ei} = 0$, hence $G_{eie} = 0$. Likewise, blocking of the inhibitory synapses also leads to $G_{eie} = 0$. Substituting $G_{eie} = 0$ in (50) and using the fact that $\tilde{h}(s) = \alpha\beta/(s+\alpha)(s+\beta)$ shows the reduction.

7.23 The current flow through a receptor is modeled by its efficacy. Since the efficacy of the inhibitory ($GABA_A$) receptors are modeled by μ_{ie}, benzodiazepine causes an increase in μ_{ie}.

7.24 The amplitude and frequency of the alpha rhythm decrease.

7.25 Yes!

Chapter 8

8.1 Although regional flow differences may be involved, there are intrinsic differences in neuronal vulnerability for hypoxic incidents. This has been found from experiments in slices. What the mechanisms involved in this intrinsic differential sensitivity are, is not exactly known.

8.2 –

8.3 Benzodiazepines, barbiturates and alcohol also interact with the $GABA_A$ at the binding sites formed by the presence of specific subunit subtypes. GABA receptors, therefore, come into various flavors, and are composed of five subunits from multiple subunit subtypes. Recent estimates indicate that there are about 30 subtypes of the $GABA_A$ receptor in the central nervous system, each with distinct physiological and pharmacological properties, and a characteristic expression pattern.

8.4

(a) The surface of the cell is approximately 1.26×10^{-9}m^2; the capacitance is 1.26×10^{-11}F.

(b) Assuming that $\varepsilon \ll x$, the membrane potential is given by (8.9) resulting in $V_m = -18.7$ mV. Using $Q = C.V_m$ the charge separated is $Q = 1.26 \cdot 10^{-11} \cdot 18.7 \cdot 10^{-3} = 2.35 \cdot 10^{-13}$ C.

(c) From the cell volume $W = 4/3\pi r^3 = 4.2 \cdot 10^{-15}\text{m}^3$ and assuming a realistic initial concentration $c = 120$ mM, $x = 120/3 = 40$ mM $= 40 \times 10^{-6}$ mol/m^3, we estimate that the charge transported is $x \cdot W \cdot 6 \cdot 10^{23} \cdot 1.6 \cdot 10^{-19} = 1.6 \cdot 10^{-8}$C. This results in a fraction $\varepsilon/x = 1.5 \cdot 10^{-5}$. Therefore, we were allowed to assume that $\varepsilon \ll x$.

8.5

(a) At room temperature T_r, we find using $c = 10$ mM that this is equivalent to an osmotic pressure of $ck_bT_r = 10 \cdot 10^{-3} \times 10^3 \times 6 \cdot 10^{23} \times 1.38 \cdot 10^{-23} \times 300 \approx 2.4$ kPa. Note that c mole per liter equals $c \times 10^3 \times N_a$ particles per m^3. Check that this is much larger than a typical cell can stand. If this situation occurs, the cell will be destroyed, therefore.
(b) The primary mechanism that prevents cells from reaching the Gibbs-Donnan equilibrium is the sodium-potassium pump, that for each cycle pumps 3 Na-ions out and brings 3 K-ions in, at the expensive a single molecule of ATP.

8.6

(a) Set $g_k = x$, and solve $x(1 + 1/25 + 1/2) = 0.2$, which results in $x = 0.13$. Then $g_k = 0.13$, $g_{\text{Na}} = 0.005$ and $g_{\text{Cl}} = 0.065$ S/m^2.

(b) For the resting membrane potential at $t = 0$, we use

$$V_m = \frac{E_k g_k + E_{\text{Na}} g_{\text{Na}} + E_{\text{Cl}} g_{\text{Cl}}}{g_{\text{k}} + g_{\text{Na}} + g_{\text{Cl}}}$$

and the Nernst potentials shown in Table B.2. Putting this in the equation above, we find $V_{\text{m}} = -67$ mV. For the sodium current density, it then holds that

$$I_{\text{Na}} = g_{\text{Na}} \cdot (E_{\text{Na}} - V_m) = 0.005 \cdot (54.7 + 67) \cdot 10^{-3} \approx 0.61 \text{ mA/m}^2.$$

Using the given geometry, with radius $r = 0.5$ mm, the axon surface for a unit length is given by $2\pi \cdot r \cdot 1 \approx 3 \cdot 10^{-3}$ m^2, resulting in an initial sodium current of $0.61 \cdot 3 \cdot 10^{-3} = 1.8$ µA.

(c) The total sodium charge inside the axon (per unit length) is given by

$$V \cdot e \cdot N_a \cdot [Na]_{\text{in}} = \pi r^2 \cdot 1.6 \cdot 10^{-19} \cdot 6 \cdot 10^{23} \cdot 50 \cdot 10^{-3} = 0.0038 \text{ C}.$$

Table B.2 Nernst potentials

$E_{\text{K}} = -75.3$ mV
$E_{\text{Na}} = 54.7$ mV
$E_{\text{Cl}} = -59.8$ mV

(d) Assuming an ion current of 1.8 μA, it takes about 70 minutes to reach an intracellular sodium concentration of 100 mmol/l.

8.7 Initially, chloride will move in the direction of its concentration gradient, adding x mmol of chloride to the left compartment, accompanied by an amount of potassium with the same value x. In equilibrium, it must hold that the Nernst potentials are the same, i.e.

$$\frac{[Cl^-]_i + x}{[Cl^-]_o - x} = \frac{[K^+]_o - x}{[K^+]_i + x}$$

Solving results in $x = 30$ mmol/l. In this situation, the osmolality of the right compartment has now decreased from 300 to 240 mM, and the left compartment has an increase in osmolality from 200 to 260 mM, resulting in a difference of 20 mM. Recall that the negative charge is carried by large macromolecules, that do not significantly affect the osmolality of the solution.

8.8 At GD equilibrium it holds that

$$\frac{[Na^+]_i}{[Na^+]_e} = \frac{[K^+]_i}{[K^+]_e} = \frac{[Cl^-]_e}{[Cl^-]_i}. \tag{B.1}$$

We also have (electroneutrality) for the intracellular (i) and extracellular (e) compartment

$$[Na^+]_j + [K^+]_j = [A^-]_j + [Cl^-]_j \tag{B.2}$$

with $j = i, e$ for the intracellular and extracellular solution, respectively. From (B.1), we set $[K^+]_i = \frac{[Na^+]_i[K^+]_e}{[Na^+]_e}$ and $[Cl^-]_i = \frac{[Na^+]_e[Cl^-]_e}{[Na^+]_i}$. We further use $[Cl^-]_e = [Na^+]_e + [K^+]_e - [A^-]_e$. Inserting in (B.2) we obtain

$$[Na^+]_i + \frac{[Na^+]_i[K^+]_e}{[Na^+]_e} = [A^-]_i + \frac{[Na^+]_e([Na^+]_e + [K^+]_e - [A^-]_e)}{[Na^+]_i}. \tag{B.3}$$

Multiply with $[Na^+]_i$ and set $\beta = [Na^+]_e + [K^+]_e$

$$[Na^+]_i^2(\frac{\beta}{[Na^+]_e}) - [Na^+]_i[A^-]_i - [Na^+]_e(\beta - [A^-]_e) = 0. \tag{B.4}$$

Using the abc-formula, we find (negative concentrations are not possible) that

$$[Na^+]_i = [Na^+]_e \frac{[A^-]_i + \sqrt{[A^-]_i^2 + 4\beta(\beta - [A^-]_e)}}{2\beta}. \tag{B.5}$$

Chapter 9

9.1 First, recall that stability can refer to a stable equilibrium but also to a limit cycle. The limit cycle is stable if perturbations away from the limit cycle (but still in the 'neighborhood') bring the dynamics back to the original limit cycle oscillations. Similarly, the equilibrium is stable if small perturbations bring the system back to its original equilibrium state.

You could for instance take the Morris-Lecar model with parameters in Fig. 4.19 as discussed in Sect. 4.7.2. The model has a stable equilibrium state with V = −26.59687 and w=0.1293793, but can also show periodic spiking, reflecting limit cycle behavior. In this case, the limit cycle is stable as small perturbations away from the limit cycle will return the system to the periodic oscillations. This region is not infinite, however: if the perturbation is relatively too large, the system will move away from the limit cycle.

If you take (v,w) = (−0.127, 0.133) as an initial condition, the system evolves towards a stable limit cycle; taking $(v, w) = (-0.127, 0.15)$, it will evolve towards the stable fixed point. Check also, that if you take as initial condition $(v, w) =$ (0.127, 0.145, which is slightly outside the stable limit cycle, or $(v, w) = (-0.127,\ 0.1)$, which is inside the stable limit cycle, the system will also evolve towards the limit cycle. Indeed, the limit cycle is stable, as nearby points do converge to it.

Check also that $(v,\ w) = (0.016328,\ 0.23974)$ is an unstable fixed point with eigenvalues $(0.00229 + 1.891i,\ 0.00229 - 1.891i)$ (see Table 4.1 or Fig. 4.2). For the other fixed points, you should find eigenvalues of (−0.258, −2.43) and (0.371, −1.58), corresponding to a stable node and a saddle node, respectively.

9.2 See [143]

9.3

9.4 Hint: consider temperature dependency of conduction velocities. This may modify the time delays involved in the thalamocortical model discussed in this chapter.

9.5

9.6

9.7

Chapter 10

10.1 –

10.2 –

10.3 Start with considering a single neuron that receives a pulse of current.

10.4 –

References

1. W.C. Abraham, O.D. Jones, D.L. Glanzman, Is plasticity of synapses the mechanism of long-term memory storage? NPJ Sci. Learn. **4**(1) (2019)
2. K. Aihara, G. Matsumoto, Two stable steady states in the Hodgkin-Huxley axons. Biophys. J. **41**(1), 87–89 (1983)
3. P. Alcamí, A.E. Pereda, Beyond plasticity: the dynamic impact of electrical synapses on neural circuits. Nat. Rev. Neurosci. **20**(5), 253–271 (2019)
4. Peter Ashwin, Stephen Coombes, Rachel Nicks, Mathematical Frameworks for Oscillatory Network Dynamics in Neuroscience Mathematical Frameworks for Oscillatory Network Dynamics in Neuroscience. J. Math. Neurosc. **6**(2), 1–92 (2016)
5. Arun Asok, Félix Leroy, Joseph B. Rayman, Eric R. Kandel, Molecular mechanisms of the memory trace. Trends Neurosci **42**(1), 14–22 (2019)
6. R.A.B. Badawy, R.A.L. Macdonell, S.F. Berkovic, S.J. Vogrin, G.D. Jackson, M.J. Cook, Reply: transcranial magnetic stimulation as a biomarker for epilepsy (Letter to the Editor). Brain (e19):1–4 (2017)
7. R.A.B. Badawy, G. Strigaro, R. Cantello, TMS, cortical excitability and epilepsy: the clinical impact. Epilepsy Res. **108**(2), 153–161 (2014)
8. E. Barreto, J.R. Cressman, Ion concentration dynamics as a mechanism for neuronal bursting. J. Biol. Phys. **37**(3), 361–373 (2011)
9. P.R. Bauer, A.A. De Goede, W.M. Stern, A.D. Pawley, F.A. Chowdhury, R.M. Helling, R. Bouet, S.N. Kalitzin, G.H. Visser, S.M. Sisodiya, J.C. Rothwell, M.P. Richardson, M.J.A.M. Van Putten, J.W. Sander, Long-interval intracortical inhibition as biomarker for epilepsy: A transcranial magnetic stimulation study. Brain **141**(2), (2018)
10. P.R. Bauer, S. Kalitzin, M. Zijlmans, J.W. Sander, G.H. Visser, Cortical excitability as a potential clinical marker of epilepsy: a review of the clinical application of transcranial magnetic stimulation. Int. J. Neural Syst. **24**(02), 1430001 (2014)
11. H. Berger, Über das Elektroenkephalogramm des Menschen. Archiv für Psychiatrie und Nervenkrankheiten **87**, 527–570 (1929)
12. G.S. Bhumbra, B.A. Bannatyne, M. Watanabe, A.J. Todd, D.J. Maxwell, M. Beato, The recurrent case for the renshaw cell. J. Neurosci. **34**(38), 12919–12932 (2014)
13. M. Bialer, H.S. White, Key factors in the discovery and development of new antiepileptic drugs. Nat. Rev. Drug Discov. **9**(1), 68–82 (2010)
14. H. Bolay, Y. Gürsoy-Özdemir, Y. Sara, R. Onur, A. Can, T. Dalkara, Persistent defect in transmitter release and synapsin phosphorylation in cerebral cortex after transient moderate ischemic injury. Stroke **33**(5), 1369–1375 (2002)
15. J.M. Bower, D. Beeman, *The Book of Genesis. Exploring Realistic Models with the General Neural Simulation System* (Springer/Elos, 2003)

M. J. A. M. van Putten, *Dynamics of Neural Networks*,
https://doi.org/10.1007/978-3-662-61184-5

16. W.E. Boyce, R.C. DiPrima, D.B. Meade, *Elementary Differential Equations*, 11th edn. (Wiley, New York, 2017)
17. R.P. Brenner, Is it status? Epilepsia **43**(Suppl 3), 103–113 (2002)
18. C.D. Brisson, R.D. Andrew, A neuronal population in hypothalamus that dramatically resists acute ischemic injury compared to neocortex. J. Neurophysiol. **108**, 419–430 (2012). https://doi.org/10.1152/jn.00090.2012
19. G. Buzsáki, *Rhythms of the Brain* (Oxford University Press, Oxford, 2006)
20. G. Buzsáki, Costas a Anastassiou, and Christof Koch. The origin of extracellular fields and currents–EEG, ECoG, LFP and spikes. Nat. Rev. Neurosci. **13**(6), 407–420 (2012)
21. C.C. Canavier, Phase-resetting as a tool of information transmission. Curr. Opin. Neurobiol. **31**, 206–213 (2015)
22. C.C. Canavier, F.G. Kazanci, A.A. Prinz, Phase resetting curves allow for simple and accurate prediction of robust N:1 phase locking for strongly coupled neural oscillators. Biophys. J. **97**(1), 59–73 (2009)
23. R. Chen, S. Dc, J. Weston, N. Sj, Transcranial magnetic stimulation for the treatment of epilepsy (Review). Cochrane Database Syst. Rev. **8**, 1–35 (2016)
24. V. Conde, L. Tomasevic, I. Akopian, K. Stanek, G.B. Saturnino, A. Thielscher, T.O. Bergmann, H.R. Siebner, The non-transcranial TMS-evoked potential is an inherent source of ambiguity in TMS-EEG studies. NeuroImage **185**, 300–312 (2019)
25. C. Conte, R. Lee, M. Sarkar, D. Terman, A mathematical model of recurrent spreading depolarizations. **203–217**, (2018)
26. D.W. Cope, G. Di Giovanni, S.J. Fyson, G. Orbán, C. Adam, M.L. Lorincz, T.M. Gould, D.A. Carter, G. Di Giovanni, A.C. Errington, V. Crunelli, L.L. Magor, S.J. Fyson, G. Orbán, A.C. Errington, M.L. Lorincz, T.M. Gould, D.A. Carter, V. Crunelli, Enhanced tonic GABAA inhibition in typical absence epilepsy. Nat. Med. **15**(12), 1392–1398 (2009)
27. A. Czaplinski, A.J. Steck, P. Fuhr, Ice pack test for myasthenia gravis. A simple, noninvasive and safe diagnostic method. J. Neurol. **250**(7), 883–884 (2003)
28. M.A. Dahlem, Mathematical modeling of human cortical spreading depression, in *Neurobiological Basis of Migraine*, eds. by T. Dalkara, M.A. Moskowitz, Chap. 17 (Wiley, New York, 2017)
29. M.A. Dahlem, T.M. Isele, Transient localized wave patterns and their application to migraine. J. Math. Neurosci. **3**(1), 1–28 (2013)
30. G. Darmani, U. Ziemann, Pharmacophysiology of TMS-evoked EEG potentials: a mini-review. Brain Stimul. **12**(3), 829–831 (2019)
31. D. Daugherty, T. Roque-Urrea, J. Urrea-Roque, J. Troyer, S. Wirkus, M.A. Porter, Mathematical models of bipolar disorder. Commun. Nonlinear Sci. Numer. Simul. **14**(7), 2897–2908 (2009)
32. A.A. de Goede, E.M. ter Braack, M.J.A.M. van Putten, Single and paired pulse transcranial magnetic stimulation in drug naïve epilepsy. Clin. Neurophysiol. **127**(9) (2016)
33. C.C. de Vos, L. Melching, J. van Schoonhoven, J.J. Ardesch, A.W. de Weerd, H.C.E. van Lambalgen, M.J.A.M. van Putten, Predicting success of vagus nerve stimulation (VNS) from interictal EEG. Seizure : J. Br. Epilepsy Assoc. **20**(7), 541–545 (2011)
34. C.C. de Vos, K. Meier, P.B. Zaalberg, H.J.A. Nijhuis, W. Duyvendak, J. Vesper, T.P. Enggaard, M.W.P.M. Lenders, Spinal cord stimulation in patients with painful diabetic neuropathy: a multicentre randomized clinical trial. Pain **155**(11), 2426–2431 (2014)
35. K. Dijkstra, J. Hofmeijer, S.A. van Gils, M.J.A.M. van Putten, A biophysical model for cytotoxic cell swelling. J. Neurosci. **36**(47), 11881–11890 (2016)
36. S.A. Dragly, M.H. Mobarhan, A.V. Solbrå, S. Tennøe, A. Hafreager, A. Malthe-Sørenssen, M. Fyhn, T. Hafting, G.T. Einevoll, Neuronify: An educational simulator for neural circuits. eNeuro **4**(2), 1–13 (2017)
37. J.P. Dreier, T. Isele, C. Reiffurth, N. Offenhauser, S. Kirov, M. Dahlem, O. Herreras, Is spreading depolarization characterized by an abrupt, massive release of gibbs free energy from the human brain cortex? Neurosci. Rev. J. Bringing Neurobiol. Neurol. Psychiatry **19**(1), 25–42 (2013)

38. J.P. Dreier, C.L. Lemale, V. Kola, A. Friedman, K. Schoknecht, Spreading depolarization is not an epiphenomenon but the principal mechanism of the cytotoxic edema in various gray matter structures of the brain during stroke. Neuropharmacology **134**, 189–207 (2018)
39. W. van Drongelen. Modeling neural activity. ISRN Biomath. (2013)
40. T.L. Eissa, K. Dijkstra, C. Brune, R.G. Emerson, M.J.A.M. van Putten, R.R. Goodman, G.M. McKhann, C.A. Schevon, W. van Drongelen, S.A. van Gils. Cross-scale effects of neural interactions during human neocortical seizure activity, in *Proceedings of the National Academy of Sciences* (2017), p. 201702490
41. J. Epsztein, Y. Ben-Ari, A. Represa, V. Crépel, Late-onset epileptogenesis and seizure genesis: lessons from models of cerebral ischemia. Neuroscientist **14**(1), 78–90 (2008)
42. D. Purves et al., (eds.) *Neuroscience* (Sinauer Associates Inc., 2004)
43. F.L. Da Silva, W. Blanes, S.N. Kalitzin, J. Parra, P. Suffczynski, D.N. Velis, Epilepsies as dynamical diseases of brain systems: basic models of the transition between normal and epileptic activity. Epilepsia **44**(12 SUPPL.), 72–83 (2003)
44. D. Fan, S. Liu, Q. Wang, Stimulus-induced epileptic spike-wave discharges in thalamocortical model with disinhibition. Sci. Rep. **6**(37703), 1–21 (2016)
45. J. Fell, P. Klaver, K. Lehnertz, T. Grunwald, C. Schaller, C.E. Elger, G. Fernández, Human memory formation is accompanied by rhinal-hippocampul coupling and decoupling. Nat. Neurosci. **4**(12), 1259–1264 (2001)
46. R. Fisher, V. Salanova, T. Witt, R. Worth, T. Henry, R. Gross, K. Oommen, I. Osorio, J. Nazzaro, D. Labar, M. Kaplitt, M. Sperling, E. Sandok, J. Neal, A. Handforth, J. Stern, A. DeSalles, S. Chung, A. Shetter, D. Bergen, R. Bakay, J. Henderson, J. French, G. Baltuch, W. Rosenfeld, A. Youkilis, W. Marks, P. Garcia, N. Barbaro, N. Fountain, C. Bazil, R. Goodman, G. McKhann, K. Babu Krishnamurthy, S. Papavassiliou, C. Epstein, J. Pollard, L. Tonder, J. Grebin, R. Coffey, N. Graves, SANTE study group. Electrical stimulation of the anterior nucleus of thalamus for treatment of refractory epilepsy. Epilepsia **51**(5), 899–908 (2010)
47. R.S. Fisher, C. Acevedo, A. Arzimanoglou, A. Bogacz, J.H. Cross, C.E. Elger, J. Engel, L. Forsgren, J. French, M. Glynn, D.C. Hesdorffer, B.I. Lee, G.W. Mathern, S.L. Moshé, E. Perucca, I.E. Scheffer, T. Tomson, M. Watanabe, S. Wiebe, ILAE official report: A practical clinical definition of epilepsy. Epilepsia **55**(4), 475–482 (2014)
48. R.S. Fisher, D.E. Blum, B. DiVentura, J. Vannest, J.D. Hixson, R. Moss, S.T. Herman, B.E. Fureman, J.A. French, Seizure diaries for clinical research and practice: limitations and future prospects. Epilepsy Behav. E&B **24**(3), 304–310 (2012)
49. A.A. De Goede, E.M. Braack, M.J.A.M. Van Putten, Single and paired pulse transcranial magnetic stimulation in drug naive epilepsy. Clinical Neurophysiol. **127**(9), 3140–3155 (2016)
50. J. Hebbink, H. Meijer, G. Huiskamp, S. van Gils, F. Leijten, Phenomenological network models: Lessons for epilepsy surgery. Epilepsia **58**(10), e147–e151 (2017)
51. R. Hindriks, H.G.E. Meijer, S.A. van Gils, M.J.A.M. van Putten, Phase-locking of epileptic spikes to ongoing delta oscillations in non-convulsive status epilepticus. Frontiers Syst. Neurosci. **7**, (2013)
52. A.L. Hodgkin, A.F. Huxley, A quantitative description of membrane current and its application to conduction and excitation in nerve. J. Physiol. **117**, 500–544 (1952)
53. J. Hofmeijer, M.C. Tjepkema-Cloostermans, M.J.A.M. van Putten, Burst-suppression with identical bursts: A distinct EEG pattern with poor outcome in postanoxic coma. Clinical Neurophysiol. **125**(5), 947–954 (2014)
54. J. Hofmeijer, M.C. Tjepkema-Cloostermans, M.J.A.M. van Putten, Outcome prediction in postanoxic coma with electroencephalography: The sooner the better. Resuscitation **91**, e1–e2 (2015)
55. J. Hofmeijer, M.J.A.M. Van Putten, Ischemic cerebral damage: an appraisal of synaptic failure. Stroke; J. Cerebral Circ. **43**(2), 607–615 (2012)
56. N. Hübel, M.S. Hosseini-Zare, J. Žiburkus, G. Ullah, The role of glutamate in neuronal ion homeostasis: A case study of spreading depolarization. PLOS Comput. Biol. **13**(10):e1005804 (2017)

57. M.J. Hull, S.R. Soffe, D.J. Willshaw, A. Roberts, Modelling feedback excitation, pacemaker properties and sensory switching of electrically coupled brainstem neurons controlling rhythmic activity. PLoS Comput. Biol. **12**(1), 1–19 (2016)
58. C. Iadecola, The neurovascular unit coming of age: a journey through neurovascular coupling in health and disease. Neuron **96**(1), 17–42 (2017)
59. E.M. Izhikevich, *Dynamical Systems in Neuroscience* (The MIT Press, Cambridge MA, 2007)
60. E.M. Izhikevich, G.M. Edelman, Large-scale model of mammalian thalamocortical systems. Proc. Natl. Acad. Sci. U. S. A. **105**(9), 3593–3598 (2008)
61. J. Jackson, E. Jambrina, J. Li, H. Marston, F. Menzies, K. Phillips, G. Gilmour, Targeting the synapse in Alzheimer's disease. Frontiers Neurosci. **13**, 1–8 (2019)
62. T. Jacob, K.P. Lillis, Z. Wang, W. Swiercz, N. Rahmati, K.J. Staley, A proposed mechanism for spontaneous transitions between interictal and ictal activity. J. Neurosci. **39**(3), 575 (2019)
63. X. Jiang, J. Gonzalez-Martinez, E. Halgren, Coordination of human hippocampal sharpwave ripples during NREM sleep with cortical theta bursts, spindles, downstates, and upstates. J. Neurosci. Official J. Soc. Neurosci. **39**(44), 8744–8761 (2019)
64. E.R. Kandel, Y. Dudai, M.R. Mayford, The molecular and systems biology of memory. Cell **157**(1), 163–186 (2014)
65. E.R.S. Kandel, T.M. Jessell (eds.) *Principles of Neural Science* (McGraw-Hill, 2015)
66. I. Khalilov, G.L. Holmes, Y. Ben-Ari, In vitro formation of a secondary epileptogenic mirror focus by interhippocampal propagation of seizures. Nat. Neurosci. **6**(10), 1079–1085 (2003)
67. V.I. Krinskii, Y.M. Kokoz, Analysis of equations of excitable membranes-I. Reduction of the Hodgkin-Huxley equations to a second order system. Biophysics **18**(3), 533–539 (1973)
68. L. Kuhlmann, K. Lehnertz, M.P. Richardson, B. Schelter, H.P. Zaveri, *Seizure prediction-ready for a new era* (Nat. Rev, Neurol, 2018)
69. M. Lauritzen, J.P. Dreier, M. Fabricius, J. Hartings, R. Graf, A.J. Strong, Clinical relevance of cortical spreading depression in neurological disorders: migraine, malignant stroke, subarachnoid and intracranial hemorrhage, and traumatic brain injury. J. Cerebral Blood Flow Metabolism: Official J. Int. Soc. Cerebral Blood Flow and Metabolism **31**(1), 17–35 (2011)
70. A.A.P. Leao, Spreading depression of activity in the cerebral cortex. J. Neurophys. **3**(28), 359–390 (1944)
71. Y. LeCun, Y. Bengio, G. Hinton, Deep learning. Nature **13**(1), 35 (2015)
72. D.T.J. Liley, I. Bojak, Understanding the transition to seizure by modeling the epileptiform activity of general anesthetic agents. J. Clin. Neurophysiol. **22**(5), 300–313 (2005)
73. G. Malagon, T. Miki, V. Tran, L. Gomez, A. Marty, Incomplete vesicular docking limits synaptic strength under high release probability conditions. eLife **9**, 1–18 (2020)
74. F. Marten, S. Rodrigues, O. Benjamin, M.P. Richardson, J.R. Terry, Onset of polyspike complexes in a mean-field model of human electroencephalography and its application to absence epilepsy. Philos. Trans. Ser. Math. Phys. Eng. Sci **367**(1891), 1145–1161 (2009)
75. D.M. Martin, A. Wong, D.R. Kumar, C.K. Loo, Validation of the 10-Item orientation questionnaire. J. ECT **1**, (2017)
76. J.G. Milton, Epilepsy as a dynamic disease: a tutorial of the past with an eye to the future. Epilepsy Behav. **18**(1–2), 33–44 (2010)
77. R.E. Mirollo, S.H. Strogatz, Synchronization of pulse-coupled biological oscillators. SIAM J. Appl. Math. **50**(6), 1645–1662 (1990)
78. D.J. Mogul, W. van Drongelen, Electrical control of epilepsy. Ann. Rev. Biomed. Eng. **16**, 483–504 (2014)
79. D.E. Naylor, H. Liu, C.G. Wasterlain, Trafficking of GABA(A) receptors, loss of inhibition, and a mechanism for pharmacoresistance in status epilepticus. J. Neurosci. Official J. Soc. Neurosci. **25**(34), 7724–7733 (2005)
80. P. Nelson, *Biological Physics* (Life. W.H. Freeman and Company, Energy, Information, 2008)
81. S.A. Oprisan, All phase resetting curves are bimodal, but some are more bimodal than others. ISRN Comput. Biol. **1–11**, (2013)
82. L. Ostergaard, J.P. Dreier, N. Hadjikhani, S.N. Jespersen, U. Dirnagl, T. Dalkara, Neurovascular coupling during cortical spreading depolarization and -depression. Stroke **1–11**, (2015)

83. O. Paulsen, T.J. Sejnowski, Natural patterns of activity and long-term synaptic plasticity. Curr. Opin. Neurobiol. **10**(2), 172–179 (2000)
84. J.T. Paz, T.J. Davidson, E.S. Frechette, B. Delord, I. Parada, K. Peng, K. Deisseroth, J.R. Huguenard, Closed-loop optogenetic control of thalamus as a tool for interrupting seizures after cortical injury. Nat. Neurosci. **16**(1), 64–70 (2013)
85. J.T. Paz, J.R. Huguenard, Microcircuits and their interactions in epilepsy: is the focus out of focus? Nat. Neurosci. **18**(3), 351–359 (2015)
86. O.V. Popovych, C. Hauptmann, P.A. Tass, Control of neuronal synchrony by nonlinear delayed feedback. Biol. Cybern. **95**(1), 69–85 (2006)
87. J.C.M. Pottkämper, J. Hofmeijer, J.A. van Waarde, M.J.A.M. van Putten, The postictal state-what do we know? Epilepsia **61**(6), 1045–1061 (2020)
88. I. Premoli, A. Biondi, S. Carlesso, D. Rivolta, M.P. Richardson, Lamotrigine and levetiracetam exert a similar modulation of TMS-evoked EEG potentials. Epilepsia **1–9**, (2016)
89. I. Premoli, D. Rivolta, S. Espenhahn, N. Castellanos, P. Belardinelli, U. Ziemann, F. Müller-Dahlhaus, Characterization of GABAB-receptor mediated neurotransmission in the human cortex by paired-pulse TMS-EEG. NeuroImage **103C**, 152–162 (2014)
90. M. Van Putten, M. Padberg, In vivo analysis of end-plate noise of human extensor digitorum brevis muscle after intramuscularly injected botulinum toxin type A. Muscle Nerve **26**, 784–790 (2002)
91. M.J.A.M Van Putten, J. Hofmeijer, Invited review. EEG monitoring in cerebral ischemia: basic concepts and clinical applications. J. Clinical Neurophysiol. **33**(3), 203–210 (2016)
92. Q. Qiu, B. Zhou, P. Wang, L. He, Y. Xiao, Y. Zhenyu, M. Zhan, Origin of amplitude synchronization in coupled nonidentical oscillators. Phys. Rev. E **101**(2) (2020)
93. B.J. Ruijter, M.J.A.M. van Putten, J. Hofmeijer, Generalized epileptiform discharges in postanoxic encephalopathy: Quantitative characterization in relation to outcome. Epilepsia **56**(11), 1845–1854 (2015). https://doi.org/10.1111/epi.13202
94. B.J. Ruijter, J. Hofmeijer, H.G.E. Meijer, M.J.A.M. van Putten, Synaptic damage underlies EEG abnormalities in postanoxic encephalopathy: a computational study. Clinical Neurophysiol. **128**(9), 1682–1695 (2017)
95. M. Saravi, A procedure for solving some second-order linear ordinary differential equations. Appl. Math. Lett. **25**(3), 408–411 (2012)
96. G.M. Shepherd (ed.), *The Synaptic Organization of the Brain* (Oxford University Press, Oxford, 2004)
97. H.Z. Shouval, S.S.H. Wang, G.M. Wittenberg, Spike timing dependent plasticity: A consequence of more fundamental learning rules. Frontiers Comput. Neurosci. **4**, 1–13 (2010)
98. H.R. Siebner, V. Conde, L. Tomasevic, A. Thielscher, T. Ole Bergmann, Distilling the essence of TMS-evoked EEG potentials (TEPs): a call for securing mechanistic specificity and experimental rigor. Brain Stimul. **12**(4), 1051–1054 (2019)
99. E. Siemkowicz, A.J. Hansen, Brain extracellular ion composition and EEG activity following 10 min ischemia in normo- and hyperglycemic rats. Stroke **12**(2), 236–240 (1981)
100. W. Singer, Neuronal synchrony: a versatile code for the definiton of relations? Neuron **24**, 111–125 (1999)
101. F.K. Skinner, H. Bazzazi, S.A. Campbell, Two-cell to N-cell heterogeneous, inhibitory networks: Precise linking of multistable and coherent properties. J. Comput. Neurosci. **18**(3), 343–352 (2005)
102. E.A. Solomon, J.E. Kragel, M.R. Sperling, A. Sharan, G. Worrell, M. Kucewicz, C.S. Inman, B. Lega, K.A. Davis, J.M. Stein, B.C. Jobst, K.A. Zaghloul, S.A. Sheth, D.S. Rizzuto, M.J. Kahana, Widespread theta synchrony and high-frequency desynchronization underlies enhanced cognition. Nat. Commun. **8**(1), (2017)
103. G.G. Somjen, Mechanisms of spreading depression and hypoxic spreading depression-like depolarization. Physiol. Rev. **81**(3), 1065–1096 (2001)
104. G.G. Somjen, *Ions in the Brain: Normal Function, Seizures and Stroke* (Oxford University Press, Oxford, 2004)

105. R.C. Sotero, *Modeling the generation of phase-amplitude coupling in cortical circuits : from detailed networks to neural mass models* (Biomed. Res, Int, 2015)
106. C.J. Stam, Y. Van Der Made, Y.A.L. Pijnenburg, Ph Scheltens, EEG synchronization in mild cognitive impairment and Alzheimer's disease. Acta Neurologica Scandinavica **108**(2), 90–96 (2003)
107. M. Stead, M. Bower, B.H. Brinkmann, K. Lee, W.R. Marsh, F.B. Meyer, B. Litt, J. Van Gompel, G. Worrell, Microseizures and the spatiotemporal scales of human partial epilepsy. Brain J. Neurol. **133**(9), 2789–2797 (2010)
108. R.A. Stefanescu, R.G. Shivakeshavan, S.S. Talathi, Computational models of epilepsy. Seizure **21**(10), 748–759 (2012)
109. I. Steuer, P.A. Guertin, Central pattern generators in the brainstem and spinal cord: an overview of basic principles, similarities and differences. Rev. Neurosci. **30**(2), 107–164 (2018)
110. P. Suffczynski, S. Kalitzin, F.H. Lopes da Silva, Dynamics of non-convulsive epileptic phenomena modeled by a bistable neuronal network. Neuroscience **126**(2), 467–484 (2004)
111. N. Suthana, Z. Haneef, J. Stern, R. Mukamel, E. Behnke, B. Knowlton, I. Fried, Memory enhancement and deep-brain stimulation of the entorhinal area. New England J. Med. **366**, 502–510 (2012)
112. J. Szentágothai, The 'module-concept' in cerebral cortex architecture. Brain Res. **95**(2–3), 475–496 (1975)
113. C. Tai, Y. Abe, R.E. Westenbroek, T. Scheuer, W.A. Catterall, Impaired excitability of somatostatin- and parvalbumin-expressing cortical interneurons in a mouse model of Dravet syndrome. Proc. Natl. Acad. Sci. **111**(30), E3139–E3148 (2014)
114. I. Tasaki, demonstration of two stable states of the nerve. J. Physiol. **148**, 306–331 (1959)
115. P.A. Tass, Desynchronization of brain rhythms with soft phase-resetting techniques. Biol. Cybern. **87**(2), 102–115 (2002)
116. P.N. Taylor, Y. Wang, G. Marc, D. Justin, M. Friederike, S. Ulrich, B. Gerold, A computational study of stimulus driven epileptic seizure abatement. PLoS ONE **9**(12), 1–26 (2014)
117. E.M. Ter Braack, A.-W.E. Koopman, M.J.A.M. van Putten, Early TMS evoked potentials in epilepsy: a pilot study. Clin. Neurophysiol. **127**(9), 3025–3032 (2016)
118. E. Thomas, S. Petrou, Network-specific mechanisms may explain the paradoxical effects of carbamazepine and phenytoin. Epilepsia **1–8**, (2013)
119. E. Thomas, S. Petrou, Network-specific mechanisms may explain the paradoxical effects of carbamazepine and phenytoin. Epilepsia **54**(7), 1195–1202 (2013)
120. M.C. Tjepkema-Cloostermans, C. da Silva Lourenço, B.J. Ruijter, S.C. Tromp, G. Drost, F.H.M. Kornips, A. Beishuizen, F.H. Bosch, J. Hofmeijer, M.J.A.M. van Putten, Outcome prediction in postanoxic coma with deep learning. Critical Care Med. 1 (2019)
121. M.C. Tjepkema-Cloostermans, R. de Carvalho, M.J.A.M. van Putten, Deep learning for detection of epileptiform discharges from scalp EEG recordings. Clin. Neurophysiol. **129**, (2018)
122. M.C. Tjepkema-Cloostermans, R. Hindriks, J. Hofmeijer, M.J.A.M. van Putten, Generalized periodic discharges after acute cerebral ischemia: reflection of selective synaptic failure? Clinical Neurophysiol. Official J. Int. Fed. Clin. Neurophysiol. **125**(2), 255–262 (2014)
123. D.M. Treiman, N.Y. Walton, C. Kendrick, A progressive sequence of electroencephalographic changes during generalized convulsive status epilepticus. Epilepsy Res. **5**(1), 49–60 (1990)
124. A.K. Tryba, E.M. Merricks, S. Lee, T. Pham, S. Cho, D.R. Nordli, T.L. Eissa, R.R. Goodman, G.M. McKhann Jr., R.G. Emerson, C.A. Schevon, W. van Drongelen, The role of paroxysmal depolarization in focal seizure activity. J. Neurophysiol. **1861–1873**, (2019)
125. P.J. Uhlhaas, W. Singer, Neural synchrony in brain disorders: relevance for cognitive dysfunctions and pathophysiology. Neuron **52**(1), 155–168 (2006)
126. G. Ullah, S.J. Schiff, Models of epilepsy. Scholarpedia **4**(7), 1409 (2009)
127. N.N. Urban, G. Barrionuevo, Induction of hebbian and non-hebbian mossy fiber long-term potentiation by distinct patterns of high-frequency stimulation. J. Neurosci. Official J. Soc. Neurosci. **16**(13), 4293–4299 (1996)
128. W. van Drongelen, H.C. Lee, M. Hereld, Z. Chen, F.P. Elsen, R.L. Stevens, Emergent epileptiform activity in neural networks with weak excitatory synapses. IEEE Trans. Neural Syst. Rehabil. Eng. **13**(2), 236–241 (2005)

129. W. van Drongelen, H.C. Lee, R.L. Stevens, M. Hereld, propagation of seizure-like activity in a model of neocortex. J. Clin. Neurophysiol. **24**(2), 182–188 (2007)
130. M.J.A.M. van Putten, *Essentials of Neurophysiology (Basic Concepts and Clinical Applications for Scientists and Engineers* (Springer, Berlin, 2009)
131. M.J.A.M. van Putten, S. Olbrich, M. Arns, Predicting sex from brain rhythms with deep learning. Sci. Rep. **8**(1), 3069 (2018)
132. M.J.A.M. van Putten, C. Jansen, M.C. Tjepkema-Cloostermans, T.M.J. Beernink, R. Koot, F. Bosch, A. Beishuizen, J. Hofmeijer, Postmortem histopathology of electroencephalography and evoked potentials in postanoxic coma. Resuscitation **134**, 26–32 (2018)
133. M.J.A.M. van Putten, M. Tjepkema-Cloostermans, J. Hofmeijer, Infraslow EEG activity modulates cortical excitability in postanoxic encephalopathy. J. Neurophysiol. **113**, 3256–3267 (2015)
134. M.J.A.M. Van Putten, J. Hofmeijer, Generalized periodic discharges: Pathophysiology and clinical considerations. Epilepsy Behav. **0–5**, (2015)
135. M.J.A.M. Van Putten, J. Hofmeijer, EEG monitoring in cerebral ischemia: Basic concepts and clinical applications. J. Clin. Neurophys. **33**(3) (2016)
136. M.J.A.M. van Putten, L. Liefaard, M. Danhof, R. Voskuyl. Four phasic response in Kainic Treated Rats
137. C.M. van Rijn, H. Krijnen, S. Menting-Hermeling, A.M.L. Coenen, Decapitation in rats: latency to unconsciousness and the 'wave of death'. PLoS One **6**(1), (2011)
138. F. Varela, J.P. Lachaux, E. Rodriguez, J. Martinerie, The brainweb: phase synchronization and large-scale integration. Nat. Rev. Neurosci. **2**(4), 229–239 (2001)
139. C. Wasterlain, D.M. Treiman (eds.), *Status epilepticus: Mechanisms and Management* (The MIT Press, Cambridge, Massahusetts, 2006)
140. S. Weisdorf, J. Duun-Henriksen, M.J. Kjeldsen, F.R. Poulsen, S.W. Gangstad, T.W. Kjær, Ultra-long-term subcutaneous home monitoring of epilepsy-490 days of EEG from nine patients. Epilepsia **1–11**, (2019)
141. T. Wennekers, F. Pasemann, Generalized types of synchronization in networks of spiking neurons. Neurocomputing **38–40**, 1037–1042 (2001)
142. H.R. Wilson, J.D. Cowan, Excitatory and inhibitory interactions in localized populations of model neurons. Biophys. J. **12**(1), 1–24 (1972)
143. B.-J. Zandt, S. Visser, M.J.A.M. van Putten, B.T. Haken, A neural mass model based on single cell dynamics to model pathophysiology. J. Comput. Neurosci.**37**(3) (2014)
144. B.-J. Zandt, T. Stigen, B.T. Haken, T. Netoff, M.J.A.M. van Putten, Single neuron dynamics during experimentally induced anoxic depolarization. J. Neurophysiol. **110**(7), 1469–1475 (2013)
145. B.-J. Zandt, B.T. Haken, J.G. van Dijk, M.J.A.M. van Putten, Neural dynamics during Anoxia and the "Wave of Death". PLoS ONE **6**(7), (2011)
146. B.-J. Zandt, B.T. Haken, M.J.A.M. van Putten, Diffusing substances during spreading depolarization: analytical expressions for propagation speed, triggering, and concentration time courses. J. Neurosci. Official J. Soc. Neurosci. **33**(14), 5915–5923 (2013)
147. B.-J. Zandt, B.T. Haken, M.J.A.M. Van Putten, M.A. Dahlem, How does spreading depression spread? Physiology and modeling. Rev. Neurosci. **26**(2), 183–198 (2015)

129. W. van Drongelen, H.C. Lee, R.E. Stevens, M. Hereld, propagation of seizure-like activity in a model of neocortex. J. Clin. Neurophysiol. 24(2), 182–188 (2007)
130. M.J.A.M. van Putten, Essentials of Neurophysiology. Basic Concepts and Clinical Applications for Scientists and Engineers (Springer, Berlin, 2009)
131. M.J.A.M. van Putten, S. Olbrich, M. Arns, Predicting sex from brain rhythms with deep learning. Sci. Rep. 8(1), 3069 (2018)
132. M.J.A.M. van Putten, [illegible], M.C. Tjepkema-Cloostermans, T.M.J. Beernink, R. Koot, [illegible]

Index

M. J. A. M. van Putten, *Dynamics of Neural Networks*,
https://doi.org/10.1007/978-3-662-61184-5

MIX
Papier aus verantwortungsvollen Quellen
Paper from responsible sources
FSC® C105338

If you have any concerns about our products,
you can contact us on
ProductSafety@springernature.com

In case Publisher is established outside the EU,
the EU authorized representative is:
Springer Nature Customer Service Center GmbH
Europaplatz 3, 69115 Heidelberg, Germany

Printed by Libri Plureos GmbH
in Hamburg, Germany